Progress in Molecular and Subcellular Biology

Series Editors: W.E.G. Müller (Managing Editor), Ph. Jeanteur,
I. Kostovic, Y. Kuchino, A. Macieira-Coelho, R. E. Rhoads

35

Volumes Published in the Series

Philippe Jeanteur (Ed.)

RNA Trafficking and Nuclear Structure Dynamics

With 28 Figures, 8 in Color

 Springer

Professor Dr. PHILIPPE JEANTEUR
Institute of Molecular Genetics
of Montpellier
BP 5051
1919 Route de Mende
34293 Montpellier Codex 05
France

ISSN 0079-6484
ISBN 978-3-540-74265-4   e-ISBN 978-3-540-74266-1

Library of Congress Control Number: 2007935103

Springer-Verlag is a part of Springer Science+Business Media
springer.com

© Springer-Verlag Berlin Heidelberg 2004, 2008

Printed on acid-free paper   39/3150ML   5 4 3 2 1 0

Preface

Recent years have witnessed an extraordinary explosion in our knowledge of the dynamics of nuclear structure. The constantly growing interest in the molecular aspects of RNA trafficking has received a dramatic stimulus from the synergistic boost of two rapidly developing technologies over recent years. On the one hand, the use of nontoxic fluorescent proteins (GFP, green fluorescent protein and its derivatives) has allowed the visualization of moving molecules in living cells due to computerized 3-D cell imaging techniques. These, in combination with FRAP (fluorescence recovery after photobleaching) methods, make it possible to follow the dynamic interactions between individual molecules. On the other hand, the development of proteomics has led to the molecular identification of proteins in motion within the nucleus.

In the face of the wealth of new observations (and reviews thereof) dealing with nuclear structure, it appeared timely to narrow the scope of this book and to restrict it to dynamic aspects, as they involve or are of relevance to RNA. It came with the realization that nuclear structure is much more dynamic than previously anticipated; this is probably the major general message of this volume.

The first three chapters deal with the structural organization of different subnuclear compartments. Nuclear compartments, unlike those in the cytoplasm, are not delimited by surrounding membranes. The chapter by Platani and Lamond surveys the chromatin compartment, the nucleolus and perinucleolar compartment, Cajal bodies and gems, the speckles containing splicing factors, as well as the PML bodies characteristic of promyelocytic leukemia.

The second chapter by Raška concentrates on the structure – function relationship of the nucleolus, in a search for active ribosomal genes. He first reviews the results obtained by a combination of electron microscopy, cytochemistry and immunocytochemistry approaches to refine nucleolar structure. He then goes on with the mapping of its DNA and RNA moieties by in situ hybridization together with its protein components by immunocytochemistry.

The next chapter by Nykamp and Swanson touches on RNA-mediated pathogenesis by presenting an original view of diseases involving trinucleotide expansion. They argue that, in addition to the protein being altered by the presence of unusual glutamic acid repeats, RNA can have an intrinsic toxic effect due to the (CUG)n expansions themselves.

The next four chapters impinge on more dynamic aspects of RNA trafficking. Bertrand and Bordonné concentrate on small nuclear and nucleolar RNPs (snRNPs and snoRNPs) with special emphasis on the latter. They draw attention to similarities in their biogenesis and argue for the possibility of a common origin between snoRNAs and snRNAs.

The other chapters focus on mRNA trafficking toward the nuclear pore. Kiesler and Visa further the monumental work of Daneholt's lab, using as a model system the Balbiani rings of *Chironomus tentans*. They track BR RNP particles from the gene to the nuclear pore, arguing for a free diffusion process with transient interactions with nonchromatin nucleoplasmic structures. They also discuss the functional significance of nuclear retention of mRNA. The paper by Braga, Rino and Carmo-Fonseca reaches the same conclusion about a passive diffusion mechanism for nuclear mRNPs, but also suggests the possibility of additional energy-dependent reactions.

Finally, Fusco, Bertrand and Singer review the latest technologies for live cell imaging of mRNA which were pioneered in Singer's lab and have already proven so useful.

Obviously, this blooming field is still in its infancy and has yet to yield its full harvest of new mechanisms and concepts.

PHILIPPE JEANTEUR

Contents

Assembly and Traffic of Small Nuclear RNPs
Edouard Bertrand and Rémy Bordonné

Intranuclear Pre-mRNA Trafficking in an Insect Model System
Eva Kiesler and Neus Visa

Photobleaching Microscopy Reveals the Dynamics of mRNA-Binding Proteins Inside Live Cell Nuclei

José Braga, José Rino and Maria Carmo-Fonseca

Imaging of Single mRNAs in the Cytoplasm of Living Cells

Dahlene Fusco, Edouard Bertrand and Robert H. Singer

Nuclear Organisation and Subnuclear Bodies

Melpomeni Platani[1] and Angus I. Lamond[1]

1
Introduction

The cell nucleus, described by Franz Bauer in 1802, was one of the first intra-cellular structures to be identified by early microscopists. The nucleus is sur-rounded by a double membrane, of which the outer membrane is continuous with the endoplasmic reticulum, and this serves to separate the sites of gene transcription from the sites of mRNA translation in the cytoplasm. Transport of molecules between the nuclear and cytoplasmic compartments of the cell occurs via specific protein complexes located in the nuclear envelope, termed nuclear pore complexes (NPCs, see Fig. 1).

The nucleus contains the genomic information of the cell and is also the site of major metabolic activities, including DNA synthesis, transcription, pre-mRNA splicing and ribosome subunit synthesis and assembly. The genome is organised in chromosomes that are packaged into large-scale domains, termed "chromosome territories", which occupy distinct regions of the nucleus (Fig. 1; Cremer et al. 1982; Cremer and Cremer 1988, 2001). It is now recognised that much of the internal structure of the nucleus, like the cytoplasm, is highly compartmentalised. As discussed below, multiple classes of subnuclear organelles, or nuclear bodies (NBs), are formed that act as compartments for specific groups of nuclear factors. A major difference between nuclear and cytoplasmic organelles, however, is that compartments in the nucleus are formed without surrounding membranes. The localisation of nuclear factors in NBs can serve several functions, including enhancing the efficiency of reac-tions by creating a high local concentration of requisite factors involved in a common process, particularly when multiple activities must act upon a single substrate, as is the case during ribosome subunit synthesis. NBs can also partition specific factors, control access of enzymes and receptors to their substrates and may play a role in intranuclear transport and in the control of gene expression.

In this review we will discuss the organisation of the cell nucleus and, in particular, consider the structure, composition and properties of some of the

[1]Wellcome Trust Biocentre, MSI/WTB Complex, DD1 5EH, Dundee, Scotland, United Kingdom.
Tel. +44-1382-345473; Fax. +44-1382-345695, e-mail: a.i.lamond@dundee.ac.uk

Progress in Molecular and Subcellular Biology
P. Jeanteur (Ed.): RNA Trafficking and Nuclear Structure Dynamics
© Springer-Verlag Berlin Heidelberg 2003

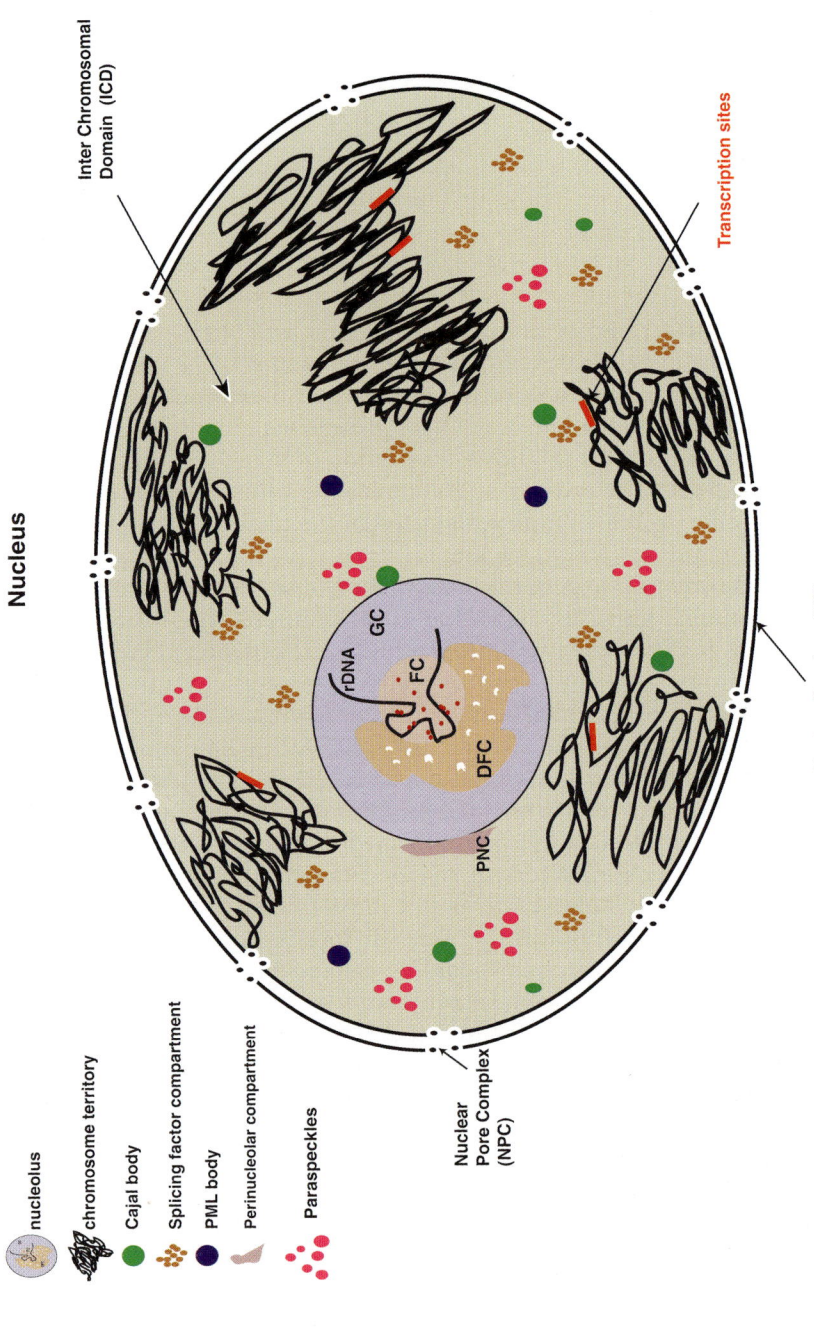

Fig. 1. Diagram of subnuclear compartments. The diagram shows the nucleus bounded by the nuclear envelope. The relative locations of chromosome territories and the principal classes of nuclear bodies that are discussed in the text are indicated

principal classes of NBs that have been characterised to date and how this relates to nuclear function.

2
Chromatin and Nuclear Compartments

The major part of the nucleoplasm is occupied by chromatin, including both decondensed euchromatin and condensed regions of transcriptionally inactive heterochromatin, separated by the interchromatin space or the "interchromo-somal domain" (ICD; Lichter et al. 1988; Zink et al. 1998; Cockell and Gasser 1999). Chromatin is composed of nucleosomes, comprising two copies of each histone protein, H2A, H2B, H3 and H4, assembled into an octamer around which 145–147 base pairs of DNA are wrapped. The repeating nucleosomes are further assembled into higher order structures, which are stabilised by the linker histone H1 (Kornberg 1977; Widom 1989; Luger et al. 1997). The strings of linked nucleosomes are helically twisted into a 10 nm fibre, which in turn is folded into a 30 nm fibre. There is also evidence from in vivo studies that interphase chromatin is actually compacted into larger scale structures, above the 30 nm chromatin fibre (Belmont et al. 1999).

Chromosome fibres follow an irregular and convoluted path in each chromosome territory. Active genes are found both at the periphery as well as within chromosome territories. The interchromatin space inside a chromosome territory is continuous with the interchromosomal domain, where RNA and proteins diffuse. There is evidence that chromatin exhibits differential compaction within separate chromosome regions. In euchromatic regions, gene activation is managed by a combination of transactivators and covalent modifications of both DNA and histones, working together with chromatin remodelling complexes (Steger et al. 1998; Strahl and Allis 2000; Jenuwein and Allis 2001; Rice and Allis 2001; Lachner and Jenuwein 2002). Heterochromatin results from functional inactivation of chromatin regions that would otherwise be transcriptionally active. They contain relatively few genes and these tend to replicate later in S-phase than active genes. Heterochromatin contains large protein complexes that can translocate along the chromatin fibre and thereby silence genes that they come in contact with (Weiler and Wakimoto 1995). The best known heterochromatin proteins are members of the HP1 family (Eissenberg and Elgin 2000).

Both heterochromatin and euchromatin can exist in the same chromosome territory and this compaction into a chromosome territory might serve a functional role. For example, it has been suggested that a possible link between chromatin condensed regions and transcriptional repression might exist (Lieb et al. 1998). This suggests that chromatin compartmentalisation may result in a greater concentration of factors at specific loci, which in turn will influence transcription and replication sites (Cockell et al. 1998).

The location of genes within the nucleus can have important consequences for both their activity and time of replication, as seen, for example, with late replicating chromatin located at the nuclear periphery (reviewed by Swedlow and Lamond 2001). Another example is provided by pericentric heterochromatin domains, which upon loss of histone modifications relocate to the nuclear periphery, lose their ability to retain HP1 and subsequently show defects in chromosome segregation during mitosis (Taddei et al. 2001).

As discussed above, individual chromosomes occupy non-overlapping territories, separated by interchromosomal domain channels (Fig. 1). RNA polymerase II transcription sites have been localised by pulse-labelling techniques and shown to occupy thousands of discrete foci distributed throughout the whole nucleus (Jackson et al. 1993; Wansink et al. 1993). In situ studies analysing specific pre-mRNAs show that they localise predominantly in round or elongated spots which colocalise with the cognate genes (Xing et al. 1993; Dirks et al. 1995). Both active and inactive genes have been localised to the periphery of chromosome territories, with non-coding sequences usually located in the interior of the territory (Kurz et al. 1996). Certain active gene loci and sites of transcription have also been mapped inside chromosome territories, suggesting that transcriptional activators can gain access inside the territory by diffusion, with nascent RNA accumulating in the interchromatin space inside and around the chromosome territories (Verschure et al. 1999; Mahy et al. 2002). The majority of pre-mRNA splicing appears to take place co-transcriptionally (Bauren and Wieslander 1994), probably involving the binding of splicing factors directly to nascent pre-mRNAs that are still tethered to their sites of transcription.

3
The Nucleolus

The most prominent and best-characterised example of a subnuclear body is the nucleolus (Fig. 2A, B). Nucleoli are the sites of ribosome subunit biogenesis and are assembled around clusters of tandemly repeated ribosomal DNA

Fig. 2. Nuclear bodies in HeLa cells. Fluorescence micrographs of HeLa cells immunolabelled with antibodies specific for the following classes of nuclear bodies. A Nucleoli, detected with an antibody against the nucleolar protein Fibrillarin (*red*). Overlay including DAPI (*blue*) staining is shown in **B**. C Cajal bodies, detected with an antibody against p80 coilin (*green*). Overlay including DAPI (*blue*) staining is shown in **D**. E Small nuclear ribonucleoproteins (snRNPs), detected with an antibody against the Sm snRNP proteins show a complex punctate pattern including diffuse nuclear staining, speckles and Cajal bodies (CBs shown by *arrowheads*). Overlay including DAPI (*blue*) staining is shown in **F**. G PML bodies, detected with an antibody against the PML protein (*green*). Overlay including the DAPI (*blue*) staining is shown in **H**. All images are maximum intensity projections of deconvolved data sets recorded using a wide field fluorescence microscope. *Scale bar* 10 μm

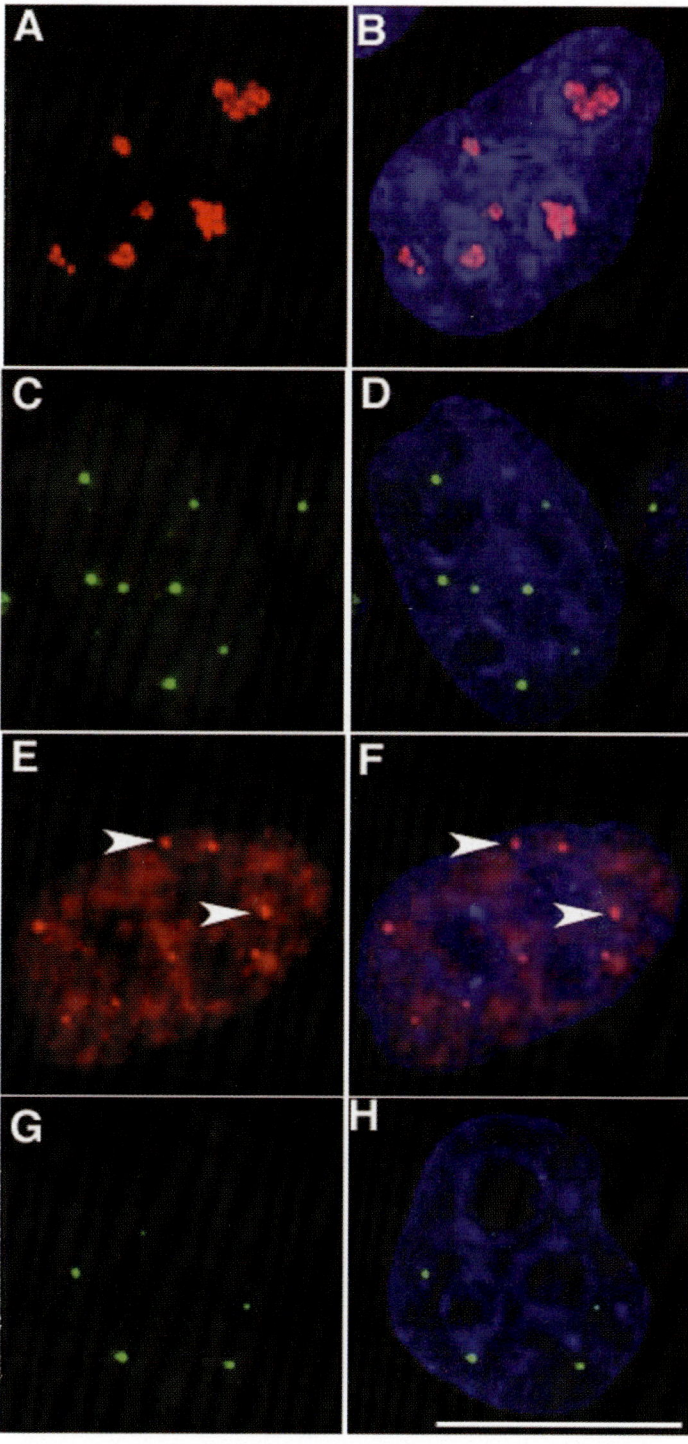

(rDNA) genes, which constitute the nucleolar organiser regions (NORs). In human cells, for example, NORs are present on chromosomes 13, 14, 15, 21, 22. The rDNA genes are transcribed by RNA Polymerase I (Scheer and Hock 1999; Carmo-Fonseca et al. 2000). When viewed by thin section electron microscopy, three distinct morphological components can be discerned in the nucleolus, i.e. fibrillar centres (FC), dense fibrillar components (DFC) and the granular component (GC). The FCs are surrounded by DFCs and the GCs emerge from these DFC regions. The rDNA genes reside at the FCs.

Not all the transcribed rDNA genes are active at any given time in the cell cycle of a HeLa cell (Jackson et al. 1998). In situ hybridisation has revealed that active rDNA genes are located at the periphery of the FC and actually extend into the DFC (Cmarko et al. 1999; Dundr and Misteli 2001). However, the exact position of the actual transcription sites is still a debated issue (Shaw et al. 1995; Scheer and Hock 1999). Studies in plant cells have demonstrated that transcription sites can reside within the DFC (Gonzales-Melendi et al. 2000). EM studies of the rDNA transcription units in intact nucleoli reveal a Christmas tree structure, with nascent pre-rRNA and ~50–100 RNA polymerase I molecules attached to the rDNA.

The 18S, 5.8S and 25/28S eukaryotic rRNAs are co-transcribed by RNA polymerase I as a pre-rRNA that is then matured and assembled with ribosomal proteins while still within the nucleolus. Following transcription, the rRNA transcripts are extensively modified in a process involving specific pre-rRNA nucleolytic cleavages, as well as both sugar and base modifications that introduce ~100 2′-O-methyl ribose and ~90 pseudouridine residues per rRNA molecule (Maden and Hughes 1997). These processing and modification events are carried out by a complex array of nucleolar factors, including predominantly the small nuclear ribonucleoprotein particles (snoRNPs; Filipowicz and Pogacic 2002). Two classes of snoRNAs, i.e. box C/D and box H/ACA snoRNAs, direct the site-specific addition of 2′-O-methyl ribose and pseudouridine formation, respectively, via complementary base pairing between snoRNAs and the rRNA substrate. Other snoRNPs, including the U3, U8, U14 and U22 snoRNPs, are required for specific nucleolytic cleavages of the pre-rRNA (Tollervey and Kiss 1997; Lewis and Tollervey 2000).

The nucleolus may also play other roles in addition to the biogenesis of ribosomes. For example, the nucleolus has been implicated in messenger RNA export or degradation, because certain polyadenylated RNAs have been detected in nucleoli (Bond and Wold 1993) and also HIV-1 mRNA has been observed trafficking through the nucleolus of virus-infected cells (Michienzi et al. 1998). It has also been implicated in the regulation of tumour suppressor protein p53 (Prives and Hall 1999) and in mediating interactions of the signal recognition particle proteins (SRP) that are involved in translation (Pederson and Politz 2000; Politz et al. 2002). In yeast, the nucleolus has been implicated in the maturation of some tRNAs, with evidence of regulation of pre-rRNA processing in response to a reduction in the activity of RNA pol III (Briand et al. 2001). The complexity of the nucleolus has also been underlined by recent proteomic studies that have documented over 300 human proteins that stably copurify with nucleoli (Anderson et al. 2002; Scherl et al. 2002).

The nucleolus is a dynamic structure that undergoes assembly and disassembly during each cell cycle. It breaks down just before the onset of mitosis and reforms in daughter nuclei during telophase, in response to the cellular requirement for new ribosome synthesis. An initial step in the breakdown of the nucleolus involves the inhibition of RNA polymerase I, probably through cdc2-cyclin B kinase inhibiting the pre-initiation complex (Heix et al. 1998; Kuhn et al. 1998), although the RNA polymerase I machinery remains associated with the NORs throughout M phase (Dundr et al. 2000; Savino et al. 2001). The mechanism of nucleologenesis is still not completely understood. The nucleoli of higher eukaryotes especially, maintain a close association with heterochromatic regions, and nucleologenesis may involve the clustering of rDNA genes via protein – protein interactions between nearby heterochromatin domains, located either on the same, or on different, chromosomes. Evidence supporting such a model comes from studies on yeast and also from the presence of heterochromatic proteins in the nucleolus (Pluta et al. 1995; Dietzel et al. 1999; Perrin et al. 1999).

4
Perinucleolar Compartment

The perinucleolar compartment (PNC) is a structure detected at the periphery of the nucleolus in some mammalian cell types. It was first identified by fluorescence in situ hybridisation studies carried out to localise human Y RNAs. Although the localisation of the Y RNAs was predominantly cytoplasmic, oligonucleotide probes against three of the four Y RNAs also hybridised to discrete structures located near the nucleolar periphery, which were correspondingly termed PNCs (Matera et al. 1995). They contain several small RNAs transcribed by RNA pol III, including the RNase MRP (RNase mitochondrial RNA processing enzyme, a nucleolar ribonucleoprotein participating in 5.8S rRNA maturation), RNase P (ribonuclease P, a nuclear endoribonuclease), and the polypyrimidine track-binding protein, hnRNP I/PTB (Matera et al. 1995; Lee et al. 1996).

A correlation has been observed between the presence of the PNC and the transformed phenotype of human cancer cells (Huang et al. 1997). The PNC dissociates at the beginning of mitosis and reforms at late telophase in daughter nuclei. When viewed in the EM it appears as an electron-dense structure consisting of multiple strands, some of which appear to be directly linked with the nucleolus and some even extend into the nucleolus (Huang et al. 1997). Incorporation of Br-UTP can be detected at the PNC, probably due to RNA polymerase II and/or RNA polymerase III, indicating that it is a site of transcription. While the biological role of the PNC is still unknown, the presence of factors needed for pre-mRNA processing suggests it may have a role in RNA metabolism (Huang et al. 1998).

Another perinucleolar structure that has been described in recent years is the Sam68 nuclear body. PNCs and SNBs share some common characteristics, but differ in their nuclear localisation (reviewed in Huang 2000).

5
Cajal Bodies, gems

Cajal bodies (CBs) were identified and described by the Spanish cytologist Ramon y Cajal at the beginning of the twentieth century (Ramon-y-Cajal 1903). They were initially termed "nucleolar accessory bodies", because of their frequent localisation at the nucleolar periphery in the neuronal cells Cajal studied. They can vary in size from 0.15–1.5 μm or larger and EM studies led to their being named "coiled bodies", because of their ultrastructural appearance as a ball of tangled threads, with 40–60 nm diameter coiled fibrillar strands (Monneron and Bernhard 1969). They were recently renamed Cajal bodies (CBs) in honour of Ramon y Cajal (Gall et al. 1999).

A human autoantigen called p80 coilin has been identified and widely used as a marker that labels CBs for both fluorescence microscopy and electron microscopy (Raska et al. 1990; Andrade et al. 1991; Raska et al. 1991). When viewed in the fluorescence microscope, coilin is usually detected in several bright foci (Fig. 2C, D). In addition, the bulk of coilin forms a diffuse nucleoplasmic pool that can cycle to and from CBs (Carmo-Fonseca et al. 1993).

In some cultured cell lines CBs were observed paired with similarly shaped nuclear structures that were termed "gems", for gemini of Cajal bodies (Liu and Dreyfuss 1996). Gems were first detected using an antibody specific for the survival of motor neuron protein, SMN (Paushkin et al. 2002). Loss of function mutations of the human gene encoding SMN results in the severe inherited muscular wasting disease called spinal muscular atrophy (SMA; Coovert et al. 1997; Lefebvre et al. 1997). SMN is part of a multi-subunit complex of gemin proteins, including gemins 1–4 that exists both in the cytoplasm and the nucleus (Liu et al. 1997; Charroux et al. 1999, 2000). The gemin complex has an important function in promoting RNP assembly, including snRNP assembly, and associates in the cytoplasm with the Sm core snRNP protein complex (see below; Pellizzoni et al. 1998; Carvalho et al. 1999; Meister et al. 2002). However, in the majority of cultured cell lines and in adult tissues it seems that both the SMN and coilin proteins precisely colocalise in the same CB structures (Young et al. 2001).

There are typically 1–10 CBs per nucleus, although some cells lack prominent CBs and their number varies between cell types and also varies during the cell cycle (Ogg and Lamond 2002). CBs are more prominent in rapidly dividing cells and their formation is transcription-dependent (Spector et al.

1992; Carmo-Fonseca et al. 1993). They disassemble during each mitosis and reassemble at the beginning of G1, following, and dependent upon, the reinitiation of transcription (Andrade et al. 1993; Ferreira et al. 1994). While there is no evidence that CBs contain DNA, there is evidence that they can associate with specific chromosomal loci, including sites of U snRNA and histone gene clusters (Jacobs et al. 1999; Shopland et al. 2001). The association of CBs with clusters of U2 snRNA genes has been shown to depend upon snRNA expression from the locus (Frey et al. 1999, 2001). CBs are able to move in the nucleoplasm, predominantly through simple diffusion (Boudonck et al. 1999; Platani et al. 2000, 2002). They also undergo both separation and joining events, including movements to and from nucleoli (Fig. 3). In HeLa cells, the rates of CB movements were found to be heterogeneous, with most rapid movements occurring in the regions of the nucleoplasm with low chromatin density (Platani et al. 2002). The tethering of CBs to specific chromosomal loci may play a role in restricting or preventing their movement at other nuclear locations.

Several lines of evidence point to CBs playing a role in the maturation of nuclear RNP complexes. In particular, it has been shown that the subsets of nuclear snRNPs and snoRNPs that localise in CBs correspond to newly assembled particles (Ogg and Lamond 2002). These RNPs subsequently leave CBs and accumulate in speckles and nucleoli, respectively (Carvalho et al. 1999; Narayanan et al. 1999; Sleeman and Lamond 1999; Sleeman 2001). CBs also contain a family of guide RNAs, termed "scaRNAs" that act via complementary base pairing interactions with snRNA substrates to target sites of 2′-O-methylribose and pseudouridine modifications (Jady and Kiss 2001). The increased prevalence of CBs in rapidly growing cells and their transcription-dependent formation may, therefore, result from CBs forming in response to a high flux of new snRNP and snoRNP formation. Consistent with this idea, overexpression of snRNP Sm proteins has been shown to induce CB formation in primary fibroblasts (Sleeman 2001).

6
Speckles – Nuclear Bodies Enriched in Splicing Factors

The pre-mRNA splicing machinery, including the small nuclear RNA-protein (snRNA-protein, or snRNP), spliceosome subunits and other non-snRNP protein splicing factors, exhibit a dynamic, punctate nuclear localisation pattern that is usually termed "speckles", or else splicing factor compartments (SFC) (Spector 1990; Spector et al. 1991). The localisation pattern of splicing factors is complex and results from the association of the splicing factors with several distinct subnuclear structures (Fig. 2E, F). Most of the punctate staining seen using anti-snRNP antibodies to detect splicing factors in the fluorescence microscope corresponds to the presence of snRNPs in bodies that are revealed in the electron microscope as interchromatin granule clusters (IGCs). A minor subset of the punctate snRNP labelling pattern results from the accumulation

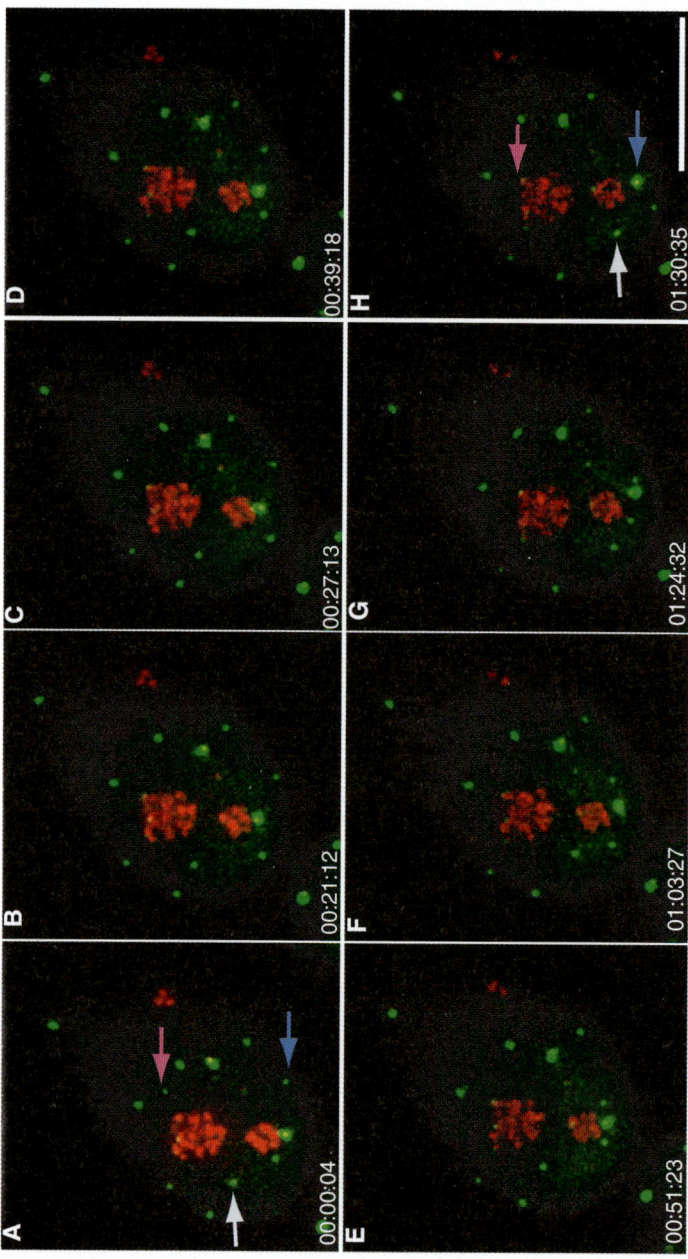

Fig. 3. In vivo analysis of three-dimensional movement of Cajal bodies and nucleoli. The figure shows representative fluorescence images from a time-lapse series of a HeLa nucleus expressing both the nucleolar protein NOPP140 (nucleolar phosphoprotein of 140 KDa) fused to the yellow fluorescent protein (YFP) and the Cajal body factor p80 coilin fused to GFP. The montage shows eight time points of dual wavelength, three-dimensional data, collected every 3 min for 1.5 h, shown as maximal intensity projections. Cajal bodies appear as bright *green* foci while the DFC nucleolar compartment is labelled in *red*. Note starting in micrograph **A** the translocation of a Cajal body through the nucleoplasm (*white arrow*), the fusion of two Cajal bodies following the movement of one through the nucleoplasm (*blue arrow*), and finally the relocation of a Cajal body from the nucleoplasm to the nucleolus (*purple arrow*), where it remains throughout the remainder of the data collection time. The final positions of each of the Cajal bodies are indicated by the same colour arrows in the final micrograph (**H**) of the time-lapse series. *Scale bar* 10 μm

of newly imported snRNPs into CBs (Fig. 2E, F, arrowheads), as discussed above. The non-snRNP protein splicing factors are not detected in CBs. Further EM studies of the intranuclear distribution of U1 and U2 snRNAs have detected a nuclear compartment termed the "interchromatin granule-associated zone". This compartment is associated with the clusters of interchromatin granules that contain U1, but not U2 snRNA, has a fibrillar texture, but does not contain DNA (Visa et al. 1993).

The IGC speckles occupy a highly variable proportion of the total nuclear volume in different cell types and can also change within a cell, both during the cell cycle and in response to changes in transcriptional activity and other metabolic perturbations. Much of the snRNP labelling in transcriptionally active cells forms a diffuse nucleoplasmic pattern that is superimposed on the speckled staining (Fig. 2E, F). At the EM level the diffuse staining includes snRNPs associated with perichromatin fibrils (PF), which are fibrillar structures 3–5 nm in diameter that correspond to nascent transcripts and often surround the IGCs (Monneron and Bernhard 1969; Fakan 1994). It is likely that most of the co-transcriptional splicing occurs in the perichromatin fibrils, i.e. in the diffuse compartment, rather than within IGC speckles. In situ hybridisation studies have localised active genes predominantly at the periphery of, rather than within, speckles (Xing et al. 1993, 1995). It has been proposed that speckles may act as storage/assembly/recycling sites of splicing factors (Caceres et al. 1994; Horowitz and Krainer 1994; Misteli 2000). Consistent with this idea, when transcription is halted, either by the use of inhibitors, or as a result of heat shock, splicing factors accumulate in speckles and the diffuse nucleoplasmic staining is lost (Spector et al. 1991; Melcak et al. 2000). In contrast, when expression of intron containing genes increases, or during viral infection when transcription levels are high, the accumulation of splicing factors in speckles is reduced and they redistribute to the diffuse nucleoplasmic transcription sites (Jimenez-Garcia et al. 1994; Huang and Spector 1996; Misteli et al. 1998). The targeting of splicing factors in vivo to sites of nascent transcription may involve a cycle of phosphorylation/dephosphorylation acting on members of the SR family of protein splicing factors (Misteli and Spector 1997; Misteli et al. 1998). Proteomic studies have been carried out on purified fractions that were enriched for IGCs and some 150 proteins detected (Mintz et al. 1999). This and other studies have shown that the speckles contain other proteins apart from splicing factors, including transcription factors (Larsson et al. 1995; Mortillaro et al. 1996; Zeng 1997) and 3′-end processing factors (Krause et al. 1994; Schul et al. 1998). Interestingly, the most recent proteomic analyses of in vitro assembled spliceosomes indicate that they may also contain transcription and 3′ processing factors, together with splicing factors, in a higher order complex (Rappsilber et al. 2002; Zhou et al. 2002).

7
Promyelocytic Leukemia Bodies

Promyelocytic leukemia (PML) bodies, also termed PODs (PML oncogenic domains), ND10 (nuclear domain 10), or Krämer bodies, provide a clear example of the link between nuclear body organisation and human disease. There can be up to 20 or more PML bodies per nucleus, although this number changes during the cell cycle, and they appear in the fluorescence microscope as roughly spherical structures with a diameter of ~0.2–1 µm (Fig. 2G, H; Dyck et al. 1994; Koken et al. 1995; Terris et al. 1995). The name PML body derives from the component protein PML, which appears to be essential for PML body formation and integrity (Wang et al. 1998). The PML protein was discovered as a consequence of the dominant oncoprotein, PML-RAR alpha that is formed between the cellular RING-finger protein PML and the retinoic acid alpha receptor as a consequence of a chromosomal translocation, and is responsible for certain forms of acute promyelocytic leukemia (APL). The PML-RAR alpha oncoprotein causes a block in myeloid cell differentiation. In cells from APL patients, the normal pattern of 10–20 PML bodies is disrupted and replaced by a pattern of micro-foci containing PML, PML-RAR alpha and the steroid receptor RXR (Dyck et al. 1994; Weis et al. 1994; Koken et al. 1995). Treatment of patients with retinoic acid, at least in some cases, can result in both remission of the transformed phenotype and in restoration of the normal pattern of PML body organisation in the nucleus. This establishes a clear link between nuclear body organisation and the disease phenotype.

It is still unclear what role PML bodies play in untransformed cells. They are sometimes associated with other nuclear bodies, such as Cajal bodies and the cleavage bodies (which are discussed below), although this is not seen in all cell types (Schul et al. 1996). Apart from the PML protein, PML bodies also contain several other known components including Sp100, retinoblastoma protein Rb, DAXX, the Bloom syndrome protein, BLM, nuclear DNA helicase II and p53 (Ruggero et al. 2000; Zhong et al. 2000; Fuchsova et al. 2002; Salomoni and Pandolfi 2002). The PML protein itself has been shown to act as a cell growth and tumour suppressor, antagonising initiation, promotion and progression of tumours of various histological origins, particularly of the lympho-hemopoietic compartment. It has also been shown to be involved in regulation of transcription (Ruggero et al. 2000; Tsukamoto et al. 2000). A number of PML body proteins, including PML and SP100, are modified by attachment of SUMO-1 (small ubiquitin-like modifier 1). While the forms of these proteins in PML bodies are conjugates with SUMO, they also are present in a diffuse nucleoplasmic pool that is not SUMO-conjugated (Sternsdorf et al. 1997; Muller et al. 1998). It has been suggested that the SUMO modification might play a crucial role in the maintenance of proper PML body formation (Ishov et al. 1999; Zhong et al. 2000). Because of the existence of ubiquitin dependent hydrolase (HAUSP) in PML bodies, it has been proposed that they might also play a role in proteasome-mediated protein degradation (Mattson et al. 2001).

PML bodies may have a role in anti-viral responses and are disrupted when cells are infected by either the human cytomegalovirus or adenovirus (Kelly et al. 1995). At the initial steps of viral infection PML bodies are detected close to viral transcription and replication sites, while accumulation of viral proteins at PML bodies later in the infection causes their disruption (Maul 1998). Recently, it has been shown that both parental and replicated herpes simplex virus type 1 (HSV-1) amplicon genomes associate with PML bodies (ND10) in live cells (Sourvinos and Everett 2002). The link between PML bodies and viral infection is strengthened by the fact that multiple PML body components are upregulated by interferons, consistent with the idea that PML bodies act as a component of the anti-viral response. It has also been suggested that PML bodies might serve as nuclear storage depots for proteins that function elsewhere in the nucleus, but which continually cycle through the PML bodies (Sterndorf et al. 1997). Recent studies have established that not only PML body components, but also PML bodies themselves are dynamic and can move in the nucleoplasm (Muratani et al. 2002). In some cases the PML body movements appeared ATP-dependent, suggesting that some form of active transport or motor-mediated mechanism may be involved.

8
Other Forms of Nuclear Bodies

Paraspeckles are punctate structures of unknown function that appear to be ubiquitous in mammalian cells. They form a pattern resembling splicing factor speckles when detected by fluorescence microscopy. However, they do not colocalise with splicing factor speckles and are separate structures (Fox et al. 2002). Three proteins that localise in paraspeckles have been identified so far, PSP1, PSP2 and p54/nrb. All three proteins have amino terminal RNA binding motifs, suggesting that the paraspeckles may play some role connected with RNA metabolism and/or transport. The PSP1 and p54/nrb proteins are closely related factors, while PSP2 is distinct and has been independently identified as a "transcriptional coactivator activator" and reported to function through association with the thyroid hormone receptor binding protein (Iwasaki et al. 2001). The p54/nrb protein has also been reported, in conjunction with other proteins, to have a role in co-transcriptional control through an interaction with steroid hormone nuclear receptors (Mathur et al. 2001), although it is also implicated in multiple other activities, including dsRNA processing (Zhang and Carmichael 2001). Interestingly, all three known paraspeckle proteins were shown to accumulate in a common cap structure at the nucleolar periphery when cells are treated with the transcription inhibitor actinomycin D (Andersen et al. 2002). Subsequent studies on cells stably expressing YFP-tagged PSP1 showed that it can continually cycle from paraspeckles to nucleoli in a transcription-dependent fashion, suggesting that paraspeckles may have a functional relationship with nucleoli (Fox 2002).

OPT bodies are distinct structures with a diameter of ~1.3 μm that spread throughout the nucleoplasm of mammalian cells in small foci. They appear during G1 phase, where they show a particular association with chromosomes 6 and 7, and disappear at the beginning of mitosis. OPT bodies contain the PTF factor (PSE-binding transcription factor), which activates the transcription of snRNA genes, together with Oct1 (also a transcription factor), RNA polymerase II and III (Grande et al. 1997), TBP and Sp1 and newly synthesised RNA. This suggests that the OPT bodies act to bring together particular genes located on specific chromosomes, forming a region where the appropriate machinery exists to regulate the expression of these genes (Pombo et al. 1998; Schul et al. 1998).

Cleavage bodies are structures found in certain mammalian cells that are enriched in factors needed for the 3′ end processing of pre-mRNAs, including CstF 64 KDa and CPSF 100 KDa proteins (Schul et al. 1996). The bulk of these factors are diffusely distributed in the nucleoplasm, but a minor fraction is also found concentrated in cleavage bodies, which are often detected either overlapping, or adjacent to, CBs. The degree of overlap between the cleavage bodies and CBs is affected by transcription. Thus, when transcription is inhibited, complete colocalisation between the cleavage bodies and CBs occurs (Schul et al. 1996). The association of cleavage bodies with CBs is also cell cycle-regulated and depends on the expression of histone genes during the cell cycle (Schul et al. 1999).

Stress-induced nuclear bodies are a small number of nuclear bodies formed in transformed cells following stress treatment, usually adjacent to nucleoli. They consist of clusters of perichromatin granules and are depots of transcripts synthesised before stress. They contain the HAP protein (hnRNP A1 interacting protein), which plays a multifunctional role in RNA metabolism and transcription, Sam68 protein, which is involved in cell cycle progression, RNA export and splicing, together with a subset of RNA processing factors. It has been proposed that stress-induced nuclear bodies might play a role in posttranscriptional regulation of gene expression in heat-shocked cells (Denegri et al. 2001, 2002).

9
Conclusion, Perspectives

Recent studies of the subcompartments in the nucleus have underlined the diverse range of nuclear bodies that exist and characterised many of the nuclear factors that concentrate in these structures. The use of in vivo imaging techniques, particularly involving the use of GFP-tagged proteins and related methods such as fluorescence recovery after photobleaching (FRAP) and fluorescence loss in photobleaching (FLIP), as well as single particle tracking (SPT), have demonstrated convincingly that the nucleus is a highly dynamic organelle. These studies indicate that proteins move in the nucleus primarily by simple diffusion (Politz et al. 1998; Daneholt 1999; Misteli 2001). Transcription factors (McNally et al. 2000; Stenoien et al. 2001), pre-mRNA splicing factors (Kruhlak et al. 2000; Phair and Misteli 2000), chromatin binding proteins (or regions of them) (Lever et al. 2000; Perche et al. 2000; Phair and Misteli 2000), DNA repair enzymes (Houtsmuller and Vermeulen 2001) and nucleolar proteins such as fibrillarin (Snaar et al. 2000), can all diffuse at different rates between sites where they are highly concentrated and the rest of the nucleoplasm. Most nuclear proteins are not fixed in one compartment, but instead can diffuse between different nuclear bodies and through the nucleoplasm (Pederson 2000). The different rates of diffusion for different factors probably reflects the different residency times of proteins at their respective high affinity binding sites (see review Carmo-Fonseca et al. 2002). It may also reflect differential transient interactions with chromatin or other nuclear substructures. The prevalence of simple diffusion eliminates the need for either signals or signal receptors to bring proteins to their high affinity sites (Dundr and Misteli 2001; Swedlow and Lamond 2001).

It is also apparent that the organisation of nuclear factors into nuclear bodies can be important for cell function and a consequence of this is the emerging link between nuclear architecture and human disease. We anticipate that the functional importance of nuclear organisation will become better understood as future studies reveal a more detailed picture of the roles played by the different classes of nuclear bodies. We also expect that in the near future proteomic studies will prove an important source of information regarding the detailed protein composition of the respective structures and this, in turn, will provide important clues and insights regarding their biological functions.

Acknowledgments. We thank members of the Lamond Lab for helpful advice and comments and Dr Alfred Vertegaal for providing the micrographs used in Fig. 2G, H. Angus Lamond is a Wellcome Trust Principal Research Fellow.

References

Andersen JS, Lyon CE, Fox A, Leung AKL, Lam YW, Steen H, Mann M, Lamond AI (2002) Directed proteomic analysis of the human nucleolus. Curr Biol 12:1–11

Andrade LE, Chan EK, Raska I, Peebles CL, Roos G, Tan EM (1991) Human autoantibody to a novel protein of the nuclear coiled body: immunological characterization and cDNA cloning of p80-coilin. J Exp Med 173:1407–1419

Andrade LEC, Tan EM, Chan EKL (1993) Immunocytochemical analysis of the coiled body in the cell-cycle and during cell proliferation. Proc Natl Acad Sci USA 90:1947–1951

Bauren G, Wieslander L (1994) Splicing of Balbiani ring 1 gene pre-mRNA occurs simultaneously with transcription. Cell 76:183–192

Belmont AS, Dietzel S, Nye AC, Strukov YG, Tumbar T (1999) Large-scale chromatin structure and function. Curr Opin Cell Biol 11:307–311

Bond VC, Wold B (1993) Nucleolar localization of myc transcripts. Mol Cell Biol 13:3221–3230

Boudonck K, Dolan L, Shaw PJ (1999) The movement of coiled bodies visualized in living plant cells by the green fluorescent protein. Mol Biol Cell 10:2297–2307

Bri JF, Navarro F, Gadal O, Thuriaux P (2001) Cross talk between tRNA and rRNA synthesis in *Saccharomyces cerevisiae*. Mol Cell Biol 21:189–195

Caceres JF, Stamm S, Helfman DM, Krainer AR (1994) Regulation of alternative splicing in vivo by overexpression of antagonistic splicing factors. Science 265:1706–1709

Carmo-Fonseca M, Ferreira J, Lamond AI (1993) Assembly of snRNP-containing coiled bodies is regulated in interphase and mitosis – evidence that the coiled body is a kinetic nuclear structure. J Cell Biol 120:841–852

Carmo-Fonseca M, Mendes-Soares L, Campos I (2000) To be or not to be in the nucleolus. Nat Cell Biol 2:E107–E112

Carmo-Fonseca M, Platani M, Swedlow JR (2002) Macromolecular mobility inside the cell nucleus. Trends Cell Biol 12:491–495

Carvalho T, Almeida F, Calapez A, Lafarga M, Berciano MT, Carmo-Fonseca M (1999) The spinal muscular atrophy disease gene product, SMN: a link between snRNP biogenesis and the Cajal (coiled) body. J Cell Biol 147:715–728

Charroux B, Pellizzoni L, Perkinson RA, Shevchenko A, Mann M, Dreyfuss G (1999) Gemin3: a novel DEAD box protein that interacts with SMN, the spinal muscular atrophy gene product, and is a component of gems. J Cell Biol Dec 13:1181–1194

Charroux B, Pellizzoni L, Perkinson RA, Yong J, Shevchenko A, Mann M, Dreyfuss G (2000) Gemin4. A novel component of the SMN complex that is found in both gems and nucleoli. J Cell Biol Mar 20:1177–1186

Cmarko D, Verschure PJ, Martin TE, Dahmus ME, Krause S, Fu XD, van Driel R, Fakan S (1999) Ultrastructural analysis of transcription and splicing in the cell nucleus after bromo-UTP microinjection. Mol Biol Cell 10:211–223

Cockell M, Gasser SM (1999) Nuclear compartments and gene regulation. Curr Opin Genet Dev 9:199–205

Cockell M, Gotta M, Palladino F, Martin SG, Gasser SM (1998) Targeting Sir proteins to sites of action: a general mechanism for regulated repression. Cold Spring Harb Symp Quant Biol 63:401–412

Coovert DD, Le TT, McAndrew PE, Strasswimmer J, Crawford TO, Mendell JR, Coulson SE, Androphy EJ, Prior TW, Burghes AH (1997) The survival motor neuron protein in spinal muscular atrophy. Hum Mol Genet 6:1205–1214

Cremer T, Cremer C, Schnieder T, Baumann H, Hens L, Kirsch-Volders M (1982) Analysis of chromosome positions in the interphase nucleus of Chinese hamster cells by laser-UV-microirradiation experiments. Hum Genet 62:201–209

Cremer T, Cremer C (1988) Centennial of Wilhelm Waldeyer's introduction of the term "chromosome" in 1888. Cytogenet Cell Genet 48:65–67

Cremer T, Cremer C (2001) Chromosome territories, nuclear architecture and gene regulation in mammalian cells. Nat Rev Genet 2:292–301

Daneholt B (1999) Pre-mRNP particles: from gene to nuclear pore. Curr Biol 9:R412–R415

Denegri M, Chiodi I, Corioni M, Cobianchi F, Riva S, Biamonti G (2001) Stress-induced nuclear bodies are sites of accumulation of pre-mRNA processing factors. Mol Biol Cell 12:3502–3514

Denegri M, Moralli D, Rocchi M, Biggiogera M, Raimondi E, Cobianchi F, De Carli L, Riva S, Biamonti G (2002) Human chromosomes 9, 12, and 15 contain the nucleation sites of stress-induced nuclear bodies. Mol Biol Cell 13:2069–2079

Dietzel S, Niemann H, Bruckner B, Maurange C, Paro R (1999) The nuclear distribution of Polycomb during *Drosophila melanogaster* development shown with a GFP fusion protein. Chromosoma 108:83–94

Dirks RW, Daniel KC, Raap AK (1995) RNAs radiate from gene to cytoplasm as revealed by fluorescence in situ hybridization. J Cell Sci 108:2565–2572

Dundr M, Misteli T (2001) Functional architecture in the cell nucleus. Biochem J 356:297–310

Dundr M, Misteli T, Olson MO (2000) The dynamics of postmitotic reassembly of the nucleolus. J Cell Biol 150:433–446

Dyck JA, Maul GG, Miller WH Jr, Chen JD, Kakizuka A, Evans RM (1994) A novel macromolecular structure is a target of the promyelocyte-retinoic acid receptor oncoprotein. Cell 76:333–343

Eissenberg JC, Elgin SC (2000) The HP1 protein family: getting a grip on chromatin. Curr Opin Genet Dev 10:204–210

Fakan S (1994) Perichromatin fibrils are in situ forms of nascent transcripts. Trends Cell Biol 4:86–90

Ferreira JA, Carmofonseca M, Lamond AI (1994) Differential interaction of splicing snrnps with coiled bodies and interchromatin granules during mitosis and assembly of daughter cell-nuclei. J Cell Biol 126:11–23

Filipowicz W, Pogacic V (2002) Biogenesis of small nucleolar ribonucleoproteins. Curr Opin Cell Biol 14:319–327

Fox AH, Lam YW, Leung AKL, Lyon CE, Andersen J, Mann M, Lamond AI (2002) Paraspeckles: a novel nuclear domain. Curr Biol 12:13–25

Frey MR, Matera AG (2001) RNA-mediated interaction of Cajal bodies and U2 snRNA genes. J Cell Biol 154:499–509

Frey MR, Bailey AD, Weiner AM, Matera AG (1999) Association of snRNA genes with coiled bodies is mediated by nascent snRNA transcripts. Curr Biol 9:126–135

Fuchsova B, Novak P, Kafkova J, Hozak P (2002) Nuclear DNA helicase II is recruited to IFN-alpha-activated transcription sites at PML nuclear bodies. J Cell Biol 158:463–473

Gall JG, Bellini M, Wu Z, Murphy C (1999) Assembly of the nuclear transcription and processing machinery: Cajal bodies (coiled bodies) and transcriptosomes. Mol Biol Cel 10:4385–4402

Gonzalez-Melendi P, Beven A, Boudonck K, Abranches R, Wells B, Dolan L, Shaw P (2000) The nucleus: a highly organized but dynamic structure. J Microsc 198:199–207

Grande MA, van der Kraan I, de Jong L, van Driel R (1997) Nuclear distribution of transcription factors in relation to sites of transcription and RNA polymerase II. J Cell Sci 110:1781–1791

Heix J, Vente A, Voit R, Budde A, Michaelidis TM, Grummt I (1998) Mitotic silencing of human rRNA synthesis: inactivation of the promoter selectivity factor SL1 by cdc2/cyclin B-mediated phosphorylation. EMBO J 17:7373–7381

Horowitz DS, Krainer AR (1994) Mechanisms for selecting 5' splice sites in mammalian pre-mRNA splicing. Trends Genet 10:100–106

Houtsmuller AB, Vermeulen W (2001) Macromolecular dynamics in living cell nuclei revealed by fluorescence redistribution after photobleaching. Histochem Cell Biol 115:13–21

Huang S (2000) Review: perinucleolar structures. J Struct Biol 129:233–240

Huang S, Deerinck TJ, Ellisman MH, Spector DL (1997) The dynamic organization of the perinucleolar compartment in the cell nucleus. J Cell Biol 137:965–974

Huang S, Deerinck TJ, Ellisman MH, Spector DL (1998) The perinucleolar compartment and transcription. J Cell Biol 143:35–47

Huang S, Spector DL (1996) Dynamic organization of pre-mRNA splicing factors. J Cell Biochem 62:191–197

Ishov AM, Sotnikov AG, Negorev D, Vladimirova OV, Neff N, Kamitani T, Yeh ET, Strauss JF 3rd, Maul GG (1999) PML is critical for ND10 formation and recruits the PML-interacting protein daxx to this nuclear structure when modified by SUMO-1. J Cell Biol 147:221–234

Iwasaki T, Chin WW, Ko L (2001) Identification and characterization of RRM-containing coactivator activator (CoAA) as TRBP-interacting protein, and its splice variant as a coactivator modulator (CoAM). J Biol Chem 276:33375–33383

Jackson DA, Hassan AB, Errington RJ, Cook PR (1993) Visualization of focal sites of transcription within human nuclei. EMBO J 12:1059–1065

Jackson DA, Iborra FJ, Manders EM, Cook PR (1998) Numbers and organization of RNA polymerases, nascent transcripts, and transcription units in HeLa nuclei. Mol Biol Cell 9:1523–1536

Jacobs EY, Frey MR, Wu W, Ingledue TC, Gebuhr TC, Gao L, Marzluff WF, Matera AG (1999) Coiled bodies preferentially associate with U4, U11, and U12 small nuclear RNA genes in interphase HeLa cells but not with U6 and U7 genes. Mol Biol Cell 10:1653–1663

Jady BE, Kiss T (2001) A small nucleolar guide RNA functions both in 2'-O-ribose methylation and pseudouridylation of the U5 spliceosomal RNA. EMBO J 20:541–551

Jenuwein T, Allis CD (2001) Translating the histone code. Science 293:1074–1080

Jimenez-Garcia LF, Segura-Valdez ML, Ochs RL, Rothblum LI, Hannan R, Spector DL (1994) Nucleologenesis: U3 snRNA-containing prenucleolar bodies move to sites of active pre-rRNA transcription after mitosis. Mol Biol Cell 5:955–966

Kelly C, van Driel R, Wilkinson GW (1995) Disruption of PML-associated nuclear bodies during human cytomegalovirus infection. J Gen Virol 76:2887–2893

Koken MH, Linares-Cruz G, Quignon F, Viron A, Chelbi-Alix MK, Sobczak-Thepot J, Juhlin L, Degos L, Calvo F, de The H (1995) The PML growth-suppressor has an altered expression in human oncogenesis. Oncogene 10:1315–1324

Kornberg RD (1977) Structure of chromatin. Annu Rev Biochem 46:931–954

Krause S, Fakan S, Weis K, Wahle E (1994) Immunodetection of poly(A) binding protein II in the cell nucleus. Exp Cell Res 214:75–82

Kruhlak MJ, Lever MA, Fischle W, Verdin E, Bazett-Jones DP, Hendzel MJ (2000) Reduced mobility of the alternate splicing factor (ASF) through the nucleoplasm and steady state speckle compartments. J Cell Biol 150:41–51

Kuhn A, Vente A, Doree M, Grummt I (1998) Mitotic phosphorylation of the TBP-containing factor SL1 represses ribosomal gene transcription. J Mol Biol 284:1–5

Kurz A, Lampel S, Nickolenko JE, Bradl J, Benner A, Zirbel RM, Cremer T, Lichter P (1996) Active and inactive genes localize preferentially in the periphery of chromosome territories. J Cell Biol 135:1195–1205

Lachner M, Jenuwein T (2002) The many faces of histone lysine methylation. Curr Opin Cell Biol 14:286–298

Larsson SH, Charlieu JP, Miyagawa K, Engelkamp D, Rassoulzadegan M, Ross A, Cuzin F, van Heyningen V, Hastie ND (1995) Subnuclear localization of WT1 in splicing or transcription factor domains is regulated by alternative splicing. Cell 81:391–401

Lee B, Matera AG, Ward DC, Craft J (1996) Association of RNase mitochondrial RNA processing enzyme with ribonuclease P in higher ordered structures in the nucleolus: a possible coordinate role in ribosome biogenesis. Proc Natl Acad Sci USA 93:11471–11476

Lefebvre S, Burlet P, Liu Q, Bertrandy S, Clermont O, Munnich A, Dreyfuss G, Melki J (1997) Correlation between severity and SMN protein level in spinal muscular atrophy. Nat Genet 16:265–269

Lever MA, Th'ng JP, Sun X, Hendzel MJ (2000) Rapid exchange of histone H1.1 on chromatin in living human cells. Nature 408:873–876

Lewis JD, Tollervey D (2000) Like attracts like: getting RNA processing together in the nucleus. Science 288:1385–1389

Lichter P, Cremer T, Borden J, Manuelidis L, Ward DC (1988) Delineation of individual human chromosomes in metaphase and interphase cells by in situ suppression hybridization using recombinant DNA libraries. Hum Genet 80:224–234

Lieb JD, Albrecht MR, Chuang PT, Meyer BJ (1998) MIX-1: an essential component of the C. elegans mitotic machinery executes X chromosome dosage compensation. Cell 92:265–277

Liu Q, Dreyfuss GA (1996) Novel nuclear-structure containing the survival of motor-neurons protein. EMBO J 15:3555–3565

Liu Q, Fischer U, Wang F, Dreyfuss G (1997) The spinal muscular atrophy disease gene product, SMN, and its associated protein SIP1 are in a complex with spliceosomal snRNP proteins. Cell 90:1013–1021

Luger K, Mader AW, Richmond RK, Sargent DF, Richmond TJ (1997) Crystal structure of the nucleosome core particle at 2.8 A resolution. Nature 389:251–260

Maden BE, Hughes JM (1997) Eukaryotic ribosomal RNA: the recent excitement in the nucleotide modification problem. Chromosoma 105:391–400

Mahy NL, Perry PE, Gilchrist S, Baldock RA, Bickmore WA (2002) Spatial organization of active and inactive genes and noncoding DNA within chromosome territories. J Cell Biol 157:579–589

Matera AG, Frey MR, Margelot K, Wolin SLA (1995) Perinucleolar compartment contains several RNA-polymerase-III transcripts as well as the polypyrimidine tract-binding protein, Hnrnp-I. J Cell Biol 129:1181–1193

Mathur M, Tucker PW, Samuels HH (2001) PSF is a novel corepressor that mediates its effect through Sin3A and the DNA binding domain of nuclear hormone receptors. Mol Cell Biol 21:2298–2311

Mattsson K, Pokrovskaja K, Kiss C, Klein G, Szekely L (2001) Proteins associated with the promyelocytic leukemia gene product (PML)-containing nuclear body move to the nucleolus upon inhibition of proteasome-dependent protein degradation. Proc Natl Acad Sci USA 98:1012–1017

Maul GG (1998) Nuclear domain 10, the site of DNA virus transcription and replication. Bioessays 20:660–667

McNally JG, Muller WG, Walker D, Wolford R, Hager GL (2000) The glucocorticoid receptor: rapid exchange with regulatory sites in living cells. Science 287:1262–1265

Meister G, Eggert C, Fischer U (2002) SMN-mediated assembly of RNPs: a complex story. Trends Cell Biol 12:472–478

Melcak I, Cermanova S, Jirsova K, Koberna K, Malinsky J, Raska I (2000) Nuclear pre-mRNA compartmentalization: trafficking of released transcripts to splicing factor reservoirs. Mol Biol Cell 11:497–510

Michienzi A, Conti L, Varano B, Prislei S, Gessani S, Bozzoni I (1998) Inhibition of human immunodeficiency virus type 1 replication by nuclear chimeric anti-HIV ribozymes in a human T lymphoblastoid cell line. Hum Gene Ther 9:621–628

Mintz PJ, Patterson SD, Neuwald AF, Spahr CS, Spector DL (1999) Purification and biochemical characterization of interchromatin granule clusters. EMBO J 18:4308–4320

Misteli T (2000) Cell biology of transcription and pre-mRNA splicing: nuclear architecture meets nuclear function. J Cell Sci 113:1841–1849

Misteli T (2001) Protein dynamics: implications for nuclear architecture and gene expression. Science 291:843–847

Misteli T, Spector DL (1997) Protein phosphorylation and the nuclear organization of pre-mRNA splicing. Trends Cell Biol 7:135–138

Misteli T, Caceres JF, Clement JQ, Krainer AR, Wilkinson MF, Spector D (1998) Serine phosphorylation of SR proteins is required for their recruitment to sites of transcription in vivo. J Cell Biol 143:297–307

Monneron A, Bernhard W (1969) Fine structural organization of the interphase nucleus in some mammalian cells. J Ultrastruct Res 27:266–288

Mortillaro MJ, Blencowe BJ, Wei X, Nakayasu H, Du L, Warren SL, Sharp PA, Berezney R (1996) A hyperphosphorylated form of the large subunit of RNA polymerase II is associated with splicing complexes and the nuclear matrix. Proc Natl Acad Sci USA 93:8253–8257

Muller S, Matunis MJ, Dejean A (1998) Conjugation with the ubiquitin-related modifier SUMO-1 regulates the partitioning of PML within the nucleus. EMBO J 17:61–70

Muratani M, Gerlich D, Janicki SM, Gebhard M, Eils R, Spector DL (2002) Metabolic-energy-dependent movement of PML bodies within the mammalian cell nucleus. Nat Cell Biol 4:106–110

Narayanan A, Speckmann W, Terns R, Terns MP (1999) Role of the box C/D motif in localization of small nucleolar RNAs to coiled bodies and nucleoli. Mol Biol Cell 10:2131–2147

Ogg SC, Lamond AI (2002) Cajal bodies and coilin – moving towards function. J Cell Biol 159:17–21

Paushkin S, Gubitz AK, Massenet S, Dreyfuss G (2002) The SMN complex, an assemblyosome of ribonucleoproteins. Curr Opin Cell Biol 14:305–312

Pederson T (2000) Diffusional protein transport within the nucleus: a message in the medium. Nat Cell Biol 2:E73–E74

Pederson T, Politz JC (2000) The nucleolus and the four ribonucleoproteins of translation. J Cell Biol 148:1091–1095

Pellizzoni L, Kataoka N, Charroux B, Dreyfuss G (1998) A novel function for SMN, the spinal muscular atrophy disease gene product, in pre-mRNA splicing. Cell 95:615–624

Perche PY, Vouc'h C, Konecny L, Souchier C, Robert-Nicoud M, Dimitrov S, Khochbin S (2000) Higher concentrations of histone macroH2A in the Barr body are correlated with higher nucleosome density. Curr Biol 10:1531–1534

Perrin L, Romby P, Laurenti P, Berenger H, Kallenbach S, Bourbon HM, Pradel J (1999) The *Drosophila* modifier of variegation modulo gene product binds specific RNA sequences at the nucleolus and interacts with DNA and chromatin in a phosphorylation-dependent manner. J Biol Chem 274:6315–6323

Phair RD, Misteli T (2000) High mobility of proteins in the mammalian cell nucleus. Nature 404:604–609

Platani M, Goldberg I, Swedlow JR, Lamond AI (2000) In vivo analysis of Cajal body movement, separation, and joining in live human cells. J Cell Biol 151:1561–1574

Platani M, Goldberg I, Lamond AI, Swedlow JR (2002) Cajal Body dynamics and association with chromatin are ATP-dependent. Nat Cell Biol 4:502–508

Pluta AF, Mackay AM, Ainsztein AM, Goldberg IG, Earnshaw WC (1995) The centromere: hub of chromosomal activities. Science 270:1591–1594

Politz JC, Browne ES, Wolf DE, Pederson T (1998) Intranuclear diffusion and hybridization state of oligonucleotides measured by fluorescence correlation spectroscopy in living cells. Proc Natl Acad Sci USA 95:6043–6048

Politz JC, Lewandowski LB, Pederson T (2002) Signal recognition particle RNA localization within the nucleolus differs from the classical sites of ribosome synthesis. J Cell Biol 159:411–418

Pombo A, Cuello P, Schul W, Yoon JB, Roeder RG, Cook PR, Murphy S (1998) Regional and temporal specialization in the nucleus: a transcriptionally-active nuclear domain rich in PTF, Oct1 and PIKA antigens associates with specific chromosomes early in the cell cycle. EMBO J 17:1768–1778

Prives C, Hall PA (1999) The p53 pathway. J Pathol 187:112–126

Ramon-y-Cajal S (1903) Un sencillo metodo de coloracion selectiva del reticulo protoplasmico y sus efectos en los diversos organos nerviosos de vertebrados e invertebrados. Trab Lab Invest Biol 2:129–221

Rappsilber J, Ryder U, Lamond AI, Mann M (2002) Large-scale proteomic analysis of the human spliceosome. Genome Res 12:1231–1245

Raska I, Ochs RL, Andrade LEC, Chan EKL, Burlingame R, Peebles C, Gruol D, Tan EM (1990) Association between the nucleolus and the coiled body. J Struct Biol 104:120–127

Raska I, Andrade LEC, Ochs RL, Chan EKL, Chang CM, Roos G, Tan EM (1991) Immunological and ultrastructural studies of the nuclear coiled body with autoimmune antibodies. Exp Cell Res 195:27–37

Rice JC, Allis CD (2001) Histone methylation versus histone acetylation: new insights into epigenetic regulation. Curr Opin Cell Biol 13:263–273

Ruggero D, Wang ZG, Pandolfi PP (2000) The puzzling multiple lives of PML and its role in the genesis of cancer. Bioessays 22:827–835

Salomoni P, Pandolfi PP (2002) The role of PML in tumor suppression. Cell 108:165–170

Savino TM, Gebrane-Younes J, de Mey J, Sibarita JB, Hernandez-Verdun D (2001) Nucleolar assembly of the rRNA processing machinery in living cells. J Cell Biol 153:1097–1110

Scheer U, Hock R (1999) Structure and function of the nucleolus. Curr Opin Cell Biol 11:385–390

Scherl A, Coute Y, Deon C, Calle A, Kindbeiter K, Sanchez JC, Greco A, Hochstrasser D, Diaz JJ (2002) Functional proteomic analysis of human nucleolus. Mol Biol Cell 13:4100–4109

Schul W, Groenhout B, Koberna K, Takagaki Y, Jenny A, Manders EM, Raska I, van Driel R, de Jong L (1996) The RNA 3' cleavage factors CstF 64 kDa and CPSF 100 kDa are concentrated in nuclear domains closely associated with coiled bodies and newly synthesized RNA. EMBO J 15:2883–2892

Schul W, de Jong L, van Driel R (1998) Nuclear neighbours: the spatial and functional organization of genes and nuclear domains. J Cell Biochem 70:159–171

Schul W, van Der Kraan I, Matera AG, van Driel R, de Jong L (1999) Nuclear domains enriched in RNA 3′-processing factors associate with coiled bodies and histone genes in a cell cycle-dependent manner. Mol Biol Cell 10:3815–3824

Shaw PJ, Highett MI, Beven AF, Jordan EG (1995) The nucleolar architecture of polymerase I transcription and processing. EMBO J 14:2896–2906

Shopl LS et al (2001) Replication-dependent histone gene expression is related to Cajal body (CB) association but does not require sustained CB contact. Mol Biol Cell 12:565–576

Sleeman JE, Ajuh PM, Lamond AI (2001) snRNP protein expression enhances the formation of Cajal bodies containing p80-coilin and SMN. J Cell Sci 114: 4407–4419

Sleeman JE, Lamond AI (1999) Newly assembled snRNPs associate with coiled bodies before speckles, suggesting a nuclear snRNP maturation pathway. Curr Biol 9:1065–1074

Snaar S, Wiesmeijer K, Jochemsen AG, Tanke HJ, Dirks RW (2000) Mutational analysis of fibrillarin and its mobility in living human cells. J Cell Biol 151:653–662

Sourvinos G, Everett RD (2002) Visualization of parental HSV-1 genomes and replication compartments in association with ND10 in live infected cells. EMBO J 21:4989–4997

Spector DL (1990) Higher-order nuclear-organization – 3-dimensional distribution of small nuclear ribonucleoprotein-particles. Proc Natl Acad Sci USA 87:147–151

Spector DL, Fu XD, Maniatis T (1991) Associations between distinct pre-messenger-RNA splicing components and the cell-nucleus. EMBO J 10:3467–3481

Spector DL, Lark G, Huang S (1992) Differences in Snrnp localization between transformed and nontransformed cells. Mol Biol Cell 3:555–569

Steger DJ, Eberharter A, John S, Grant PA, Workman JL (1998) Purified histone acetyltransferase complexes stimulate HIV-1 transcription from preassembled nucleosomal arrays. Proc Natl Acad Sci USA 95:12924–12929

Stenoien DL, Patel K, Mancini MG, Dutertre M, Smith CL, O'Malley BW, Mancini MA (2001) FRAP reveals that mobility of oestrogen receptor-alpha is ligand- and proteasome-dependent. Nat Cell Biol 3:15–23

Sternsdorf T, Grotzinger T, Jensen K, Will H (1997) Nuclear dots: actors on many stages. Immunobiology 198:307–331

Strahl BD, Allis CD (2000) The language of covalent histone modifications. Nature 403:41–45

Swedlow JR, Lamond AI (2001) Nuclear dynamics: where genes are and how they got there. Genome Biol 2:3

Taddei A, Maison C, Roche D, Almouzni G (2001) Reversible disruption of pericentric heterochromatin and centromere function by inhibiting deacetylases. Nat Cell Biol 3:114–120

Terris B, Baldin V, Dubois S, Degott C, Flejou JF, Henin D, Dejean A (1995) PML nuclear bodies are general targets for inflammation and cell proliferation. Cancer Res 55:1590–1597

Tollervey D, Kiss T (1997) Function and synthesis of small nucleolar RNAs. Curr Opin Cell Biol 9:337–342

Tsukamoto T, Hashiguchi N, Janicki SM, Tumbar T, Belmont AS, Spector DL (2000) Visualization of gene activity in living cells. Nat Cell Biol 2:871–878

Verschure PJ, van Der Kraan I, Manders EM, van Driel R (1999) Spatial relationship between transcription sites and chromosome territories. J Cell Biol 147:13–24

Visa N, Puvion-Dutilleul F, Bachellerie JP, Puvion E (1993) Intranuclear distribution of U1 and U2 snRNAs visualized by high resolution in situ hybridization: revelation of a novel compartment containing U1 but not U2 snRNA in HeLa cells. Eur J Cell Biol 60:308–321

Wang ZG, Ruggero D, Ronchetti S, Zhong S, Gaboli M, Rivi R, Pandolfi PP (1998) PML is essential for multiple apoptotic pathways. Nat Genet 20:266–272

Wansink DG, Schul W, van der Kraan I, van Steensel B, van Driel R, de Jong L (1993) Fluorescent labeling of nascent RNA reveals transcription by RNA polymerase II in domains scattered throughout the nucleus. J Cell Biol 122:283–293

Weiler KS, Wakimoto BT (1995) Heterochromatin and gene expression in Drosophila. Annu Rev Genet 29:577–605

Weis K, Rambaud S, Carvalho T, Carmo-Fonseca M, Lavau C, Jansen J, Lamond AI, Dejean A (1994) Retinoic acid regulates aberrant nuclear localization of PML-RAR alpha in acute promyelocytic leukemia cells. Cell 76:345–356

Widom J (1989) Toward a unified model of chromatin folding. Annu Rev Biophys Biophys Chem 18:365–395

Xing Y, Johnson CV, Dobner PR, Lawrence JB (1993) Higher level organization of individual gene transcription and RNA splicing. Science 259:1326–1330

Xing Y, Johnson CV, Moen PT Jr, McNeil JA, Lawrence J (1995) Nonrandom gene organization: structural arrangements of specific pre-mRNA transcription and splicing with SC-35 domains. J Cell Biol 131:1635–1647

Young PJ, Le TT, Dunckley M, Nguyen TM, Burghes AH, Morris GE (2001) Nuclear gems and Cajal (coiled) bodies in fetal tissues: nucleolar distribution of the spinal muscular atrophy protein, SMN. Exp Cell Res 265:252–261

Zeng C (1997) Dynamic relocation of transcription and splicing factors dependent upon transcriptional activity. EMBO J 16:1401–1412

Zhang Z, Carmichael GG (2001) The fate of dsRNA in the nucleus: a p54(nrb)-containing complex mediates the nuclear retention of promiscuously A-to-I edited RNAs. Cell 106:465–475

Zhong S, Salomoni P, Ronchetti S, Guo A, Ruggero D, Pandolfi PP (2000) Promyelocytic leukemia protein (PML) and Daxx participate in a novel nuclear pathway for apoptosis. J Exp Med 191:631–640

Zhou Z, Licklider LJ, Gygi SP, Reed R (2002) Comprehensive proteomic analysis of the human spliceosome. Nature 419:182–185

Zink D, Cremer T, Saffrich R, Fischer R, Trendelenburg MF, Ansorge W, Stelzer EH (1998) Structure and dynamics of human interphase chromosome territories in vivo. Hum Genet 102:241–251

Searching for Active Ribosomal Genes

Ivan Raška[1]

1
Introduction

The nucleolus is a subnuclear organelle (a membraneless nuclear compartment) in which, besides some other functions (Olson et al. 2002; Bernardi and Pandolfi 2003), the biogenesis of ribosomes takes place (e.g., Lewis et al. 1966; Busch and Smetana 1970; Goessens 1984; Hadjiolov 1985; Fakan 1986; Risueno and Medina 1986; Raška et al. 1990; Scheer and Benavente 1990; Hernandez-Verdun 1991; Thiry et al. 1991; Wachtler and Stahl 1993; Shaw and Jordan 1995; Carmo-Fonseca et al. 1996; Reeder 1999; Scheer and Hock 1999; Warner 1999; Olson et al. 2000; Fatica and Tollervey 2002). It is characterized by a high concentration of RNA (Brachet 1940; Caspersson and Schultz 1940) and thus is the most conspicuous nuclear structure seen through light and electron microscopy (LM and EM). Indeed, in metabolically active animal and plant somatic cells, the nucleolus harbors many tens, sometimes hundreds, of active ribosomal genes that are usually characterized by high levels of transcription accounting for more than 50% of the total cellular RNA production. At the same time, however, a significant fraction of nucleolar ribosomal genes, possibly in the order of magnitude of 50–95% (including the situation in yeast), are transcriptionally silent and likely exhibit the nucleosomal chromatin arrangement (Scheer 1978; Prior et al. 1983; Conconi et al. 1989; Gonzales-Melendi et al. 2001; Santoro and Grummt 2001; Sandmeier et al. 2002; Santoro et al. 2002).

The somatic cells of higher eukaryotes contain hundreds or even thousands of ribosomal genes (Hadjiolov 1985). Typically, the genes are tandem repeats and form arrays on several chromosomes. Such chromosomal gene clusters are termed nucleolus organizer regions (NORs). For instance, human diploid cells contain about 400 ribosomal genes organized in the form of several tens of head-to-tail tandem repeats at well described positions within five pairs of chromosomes (Hadjiolov 1985). During interphase, a phenomenon termed nucleolar fusion takes place during which NORs from more than one chromo-

[1]Laboratory of Gene Expression, First Faculty of Medicine, Charles University and Department of Cell Biology, Institute of Experimental Medicine, Academy of Sciences of the Czech Republic, Albertov 4, 12800, Prague 2, Czech Republic. Tel. +420-224910315; Fax. +420-224917418, e-mail: iraska@lf1.cuni.cz

Progress in Molecular and Subcellular Biology
P. Jeanteur (Ed.): RNA Trafficking and Nuclear Structure Dynamics
© Springer-Verlag Berlin Heidelberg 2003

some often participate in the formation of a given nucleolus (Anastossova-Kristeva 1977).

Each ribosomal gene unit usually consists of a transcribed sequence and an external nontranscribed spacer (Scherrer et al. 1963; Lewis et al. 1966; Liau and Perry 1969; Hadjiolov 1995; Lazdins et al. 1997; Fatica and Tollervey 2002). The nucleolar RNA polymerase I synthesizes the long precursor ribosomal RNA (pre-rRNA) that (in human cells) contains, in addition to the 18S, 5.8S and 28S rRNA sequence, long internal and external transcribed spacer sequences. The noncoding spacer sequences are removed before or during the nucleolar assembly of the 40S and 60S ribosomal subunits. It should be mentioned here that the sizes of both coding and noncoding sequences vary among species. The biogenesis of ribosomal RNA and ribosomes is a complex process which necessitates the presence of ribonucleoproteins (RNPs) containing large varieties of small nucleolar RNAs (snoRNAs; Filipowicz and Pogačic 2002) and involves, besides the mentioned cleavage of spacers, the modification of about 200 nucleosides, and the association of rRNA with ribosomal proteins and 5S rRNA. Over the past few years, we have witnessed exciting progress with regard to our molecular knowledge of ribosome biogenesis, including even the transport of ribosomes to the cytoplasm (Schäfer et al. 2003). This was mainly due to extensive proteomic research implemented recently on nucleoli (Andersen et al. 2002; Scherl et al. 2002) and the enormous flow of data particularly from the yeast model that was directly supported by the implemented genetics (Fatica and Tollervey 2002). In this sense, our knowledge about the molecular ribosome biogenesis in higher eukaryotes lags somewhat behind that in yeast (Fatica and Tollervey 2002).

In thin sections, three basic nucleolar subcompartments are seen in typical nucleoli of animal and plant somatic cells (Fig. 1). First, nucleolar electron lucent fibrillar centers (FCs) are frequently circular in shape and contain little RNA (Recher et al. 1969; Goessens 1984). There is a general consensus that these structures represent the interphasic "counterparts" of NORs, and thus contain ribosomal DNA (rDNA). However, in metabolically active and cycling cells, such as human HeLa cells, individual "NORs" unravel and generate several FCs as there are many more FCs than NORs. Second, they are nucleolar dense fibrillar components (DFCs) that usually form a rim around the FCs and sometimes protuberances into the FCs. They contain a high concentration of RNA and are, therefore, electron-dense. A consensus exists that the first steps in the pre-rRNA processing take place in this nucleolar subcompartment (e.g., Shaw and Jordan 1995). The FCs and DFCs are embedded in the mass of closely packed 15 nm granules termed nucleolar granular components (GCs) that represent the third nucleolar subcompartment. It is generally accepted that the GCs represent, at least partly, the pre-ribosomal particles and that the later steps of pre-rRNA processing take place in this nucleolar subcompartment. From the analysis of serially sectioned mammalian nucleoli, it follows that the FCs form individual focal entities (Junera et al. 1995; Koberna et al. 2002). The DFCs form a rim around the FCs, but frequently also form a sort of intranu-

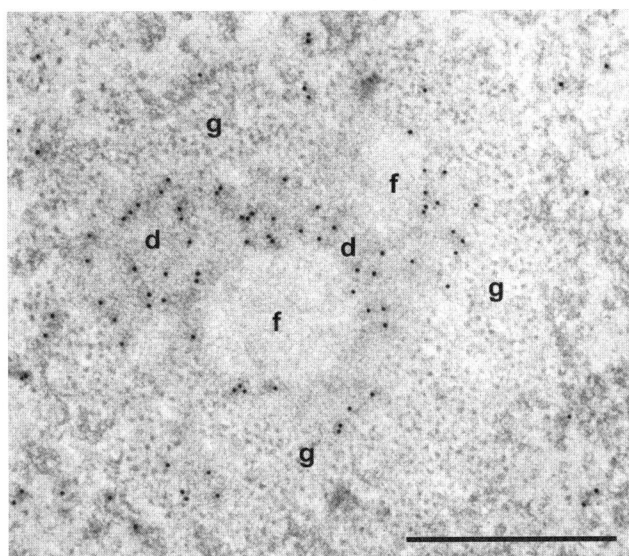

Fig. 1. EM mapping of nucleolar transcription in (resin-embedded and thin-sectioned) HeLa cells. The cell culture was placed for 5 min in the cold hypotonic buffer containing BrUTP and then incubated for 5 min in the normal medium. The incorporated BrU was revealed by means of on section labeling with the monoclonal antibody to BrU followed by the 10-nm gold adduct. Note that the nucleolar ultrastructure is well preserved and the three basic nucleolar components are clearly distinguished. Gold particles within the nucleolus are confined to the DFCs together with the DFC/FC border. *f*, *d* and *g* correspond to fibrillar centers, dense fibrillar components and granular components, respectively. *Bar* 0.2 μm. Reproduced from Koberna et al. (2002) (Fig. 3A) by kind copyright permission of the Rockefeller University Press

cleolar reticulum (Junera et al. 1995). The basic morphology of animal and plant nucleoli are compatible, but in animal cells, the GCs represent most of the nuclear area in sectioned nucleoli, whereas DFCs occupy most of the section in plant nucleoli, and in addition, the DFCs in animal cells are more electron-dense than in plants (Jordan and McGovern 1981; Shaw and Jordan 1995).

In order to provide the reader with a complete EM image of the thin-sectioned nucleolus, it is worthy of mention that much of the heterochromatin, termed the perinucleolar condensed chromatin, is associated with the periphery of nucleoli, and that the so-called nucleolar interstices, termed alternatively nucleolar vacuoles, of variable morphology as well as clumps of heterochromatin can also be seen within the nucleolar body (Busch and Smetana 1970; Smetana and Busch 1974; Goessens 1984; Hadjiolov 1985; Derenzini et al. 1990; Thiry et al. 1991). Importantly, rDNA sequences represent in mammalian cells at most a small percentage of the total intranucleolar DNA sequences (Bachellerie et al. 1977).

The production of ribosomes is essential for cell metabolism and the nucleolar morphology testifies in many aspects to the general metabolism of the cell. For instance, dormant human lymphocytes, either in smears of peripheral blood, or isolated from peripheral blood of healthy donors, frequently contain a single small nucleolus, termed ring-shaped nucleolus that exhibits just one FC (Smetana et al. 1967, 1968; Raška et al. 1983a). Stimulated human lymphocytes (e.g., by phytohemagglutinin) exhibit after 24–48 h culture several large nucleoli that exhibit many FCs (Busch and Smetana 1970; Biberfeld 1971; Raška et al. 1983b; Ochs and Smetana 1989). The nucleolar morphology has proved to be a convenient diagnostic marker in human pathology, particularly in cancers (Busch and Smetana 1970). The metabolically active and cycling mammalian somatic cells that necessitate the high production of ribosomes usually contain several large nucleoli with many tiny fibrillar centers (Busch and Smetana 1970; Koberna et al. 2002). If such cells possess nucleoli with one or only a few large fibrillar centers, it usually means that these cells are not that "happy".

The molecular organization of active ribosomal genes in the form of Christmas trees (Fig. 2) was described on spreads of nuclear contents from amphibian oocytes more than 30 years ago (Miller and Beatty 1969). Christmas trees were later described by many other groups and similar structures were also reported in other species including yeast cells (e.g., Trendelenburg 1974; Trendelenburg et al. 1974; Puvion-Duttileul et al. 1977; Scheer 1978, 1987; Franke et al. 1979; Scheer et al. 1981; Chooi and Leiby 1981; Mougey et al. 1993; Osheim et al. 1996; Dragon et al. 2002). These images showed the path of the gene, i.e. the trunk of the tree, and usually about 100 nascent pre-rRNAs radiating away in a gradient of increasing length, i.e. the branches of the tree. Even though the Christmas trees are, to a large extent, deproteinized, the associations of

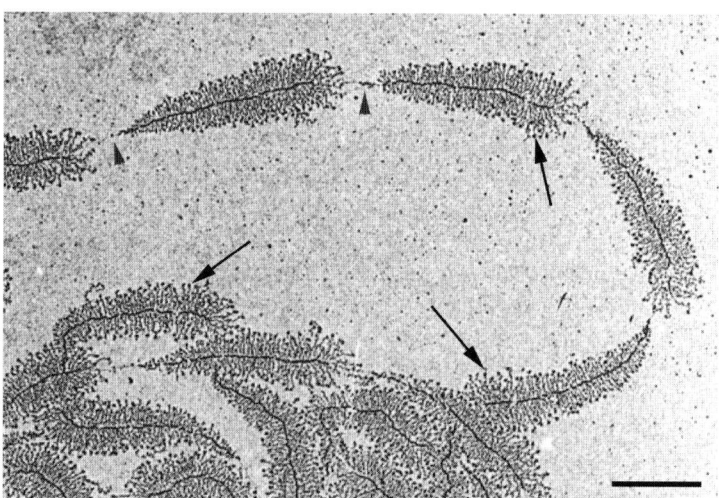

Fig. 2. Spread active ribosomal genes from Pleurodele oocytes seen in the form of the "Christmas trees". The *arrowheads* designate the rDNA, *arrows* nascent pre-rRNAs. *Bar* 1 μm. The figure was kindly provided by Professor Ulrich Scheer, University of Würzburg

RNA polymerase I, complexes involved in the early processing of pre-rRNA (RNP particle containing U3 snoRNA) and even ribosomal proteins from both small and large subunit with nascent rRNA are clearly shown (Chooi and Leiby 1981; Scheer and Benavente 1990; Mougey et al. 1993; Dragon et al. 2002), and in *Dictyostelium* and yeast, co-transcriptional cleavage has been observed in the internal transcribed spacer region, indicating that both this step, and the earlier cleavage releasing the 5′ external transcribed spacer, can take place on nascent transcripts (Grainger and Maizels 1980; Beyer, pers. comm.). Ribosomal proteins were shown to associate in the specific sequence with nascent pre-rRNA and proteins of the small ribosomal subunit were associating, within the Christmas trees, with the 18S sequence of the pre-rRNA while those from the large subunit with the 28S sequence of the pre-rRNA (Chooi and Leiby 1981). The U3 snoRNP processome was shown to correspond to the terminal knobs of the 5′end of the nascent transcript in Christmas trees (Dragon et al. 2002).

Surprisingly, the first in situ descriptions of the morphological equivalents of Christmas trees appeared only recently. Structures compatible with Christmas trees in a compacted form, respectively the intense nucleolar transcription signal compatible with the presence of compacted Christmas trees, were observed in specialized nucleoli of grasshopper oocytes (Scheer et al. 1997), in the somatic cells of *Pisum sativum* (Gonzalez-Melendi et al. 2001) and in cultured human HeLa cells (Koberna et al. 2002). The ribosomal genes in animal cells, with regard to the steady state cell metabolism, were usually reported as either fully inactive or fully active under physiological conditions. However, a reduced transcriptional activity of spread ribosomal genes was reported during the maturation of amphibian oocytes that is known to be accompanied by the lowering of the rDNA transcription activity (Scheer 1978). It is necessary to mention here that the active ribosomal genes in actively growing yeast also exhibit active genes with a lower number of nascent rRNA chains (French et al. 2002). However, there is still a large fraction of active ribosomal genes in yeast that have the form of true Christmas trees, eventually of Christmas trees with some rRNA "branches" missing (French et al. 2002).

The aim of this study is to advance a view which addresses the long-standing controversy with regard to the localization of highly active ribosomal genes within the nucleolar architecture of animal somatic cells. In other words, the question is: where are the "Christmas trees"[2] within the nucleolus? While the highly active ribosomal genes have been localized exclusively into the DFCs in plant somatic cells (e.g., Melčák et al. 1996; de Carcer and Medina 1999; Gonzalez-Melendi et al. 2001), nucleolar transcription sites in animal somatic cells have been mapped either to the FCs and the FC/DFC border (Thiry et al. 2000; Cheutin et al. 2002) or to the DFC and the DFC/FC transition zone (Granboulan and Granboulan 1965; Royal and Simard 1975; Thiry et al. 1985; Dundr and Raška 1993; Hozák et al. 1994; Raška et al. 1995; Melčák et al. 1996;

---

[2]I shall use quotation marks in the text when describing "Christmas trees" in situ

Mosgöller et al. 1998, 2001; Koberna et al. 1999, 2002; Cmarko et al. 2000; Gonzalez-Melendi et al. 2001; Staněk et al. 2001). It may seem somewhat disturbing to discuss such an "elementary" question here in the context of what has been stated above about the largely expanded molecular knowledge of the biogenesis of rRNA and ribosomes. On the contrary, this controversial issue is to be settled once and for all as the two views could not be reconciled up to now and represent an uneasiness for cell biologists working in the nucleolar field.

2
Historical Perspective

For a long time, there existed a consensus based on the high resolution autoradiography data with tritiated uridine that the transcription of ribosomal genes in mammalian cells takes places in the DFCs including the DFC/FC border (Granboulan and Granboulan 1965; Geuskens and Bernard 1966; Royal and Simard 1974; Goessens 1976; Goessens and Lepoint 1979; Thiry et al. 1985; Fakan 1986). However, the neat images of the immunolocalization of RNA polymerase I to the FCs (Scheer and Rose 1984) by the pre-embedding EM approach "moved" the sites of transcription from the DFCs to the FCs. The mapping results of RNA polymerase I were further substantiated by the postembedding approach (Scheer and Raška 1987). This represented a real schism in the nucleolar field, as, according to the results obtained, some groups stuck to the original concept, whereas others considered the FC to contain the transcription sites (Raška et al. 1995). The schism was further substantiated by a failure to demonstrate rDNA, DNA and various chromatin components within the DFCs by several groups (e.g., Thiry and Thiry-Blaise 1989; Puvion-Dutilleul et al. 1991a; Thiry et al. 1991) and also by the high resolution autoradiography findings of Thiry and Goessens (1991) who have shown that the FCs, and not the DFCs, were the major site of incorporation of tritiated uridine in cultured mammalian cells. Interestingly enough, the original findings based upon the RNA polymerase I mappings were meanwhile rectified in a gentle way. Indeed, the DFC signal due to the enzyme was smaller than that of the FCs, but significant (Raška et al. 1989; Fig. 3).

Over the last 10 years, however, the concept that the interior of the FCs harbors the active genes has changed as the transcription sites are being moved from the interior to the periphery of the FCs only (Dundr and Raška 1993). In this context, I find the term "periphery of FCs" not as suitable as "proximity of DFCs" (Raška et al. 1995). Fortunately or unfortunately, the most recent paper by Cheutin et al. (2002), based mainly on the tomographic reconstruction of the RNA polymerase I labeling, identifies once again whole FCs with the transcription sites in animal somatic cells.

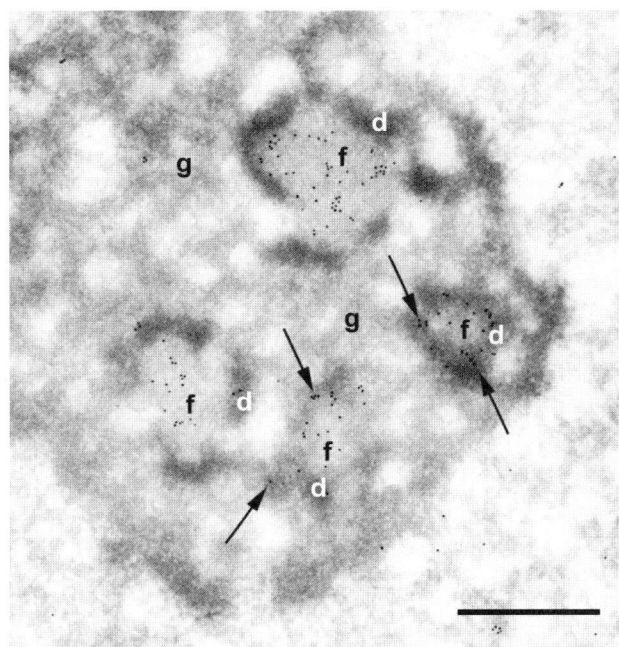

Fig. 3. Mapping of RNA polymerase I in the nucleolus of a thin cryo-sectioned HeLa cell. Several FCs are seen in the section. Note that the signal (10 nm gold particles) is seen in the FCs, but also (*arrows*) in the DFCs. *f, d* and *g* correspond to fibrillar centers, dense fibrillar components and granular components, respectively. *Bar* 0.5 µm. Reproduced from Raška et al. (1989) (Fig. 2) with kind copyright permission of Elsevier

3
The Ups and Downs of Affinity Cytochemistry: the Case of the Nucleolus

The flood of new results obtained during the last 15 years in cell biology nuclear research has, to a large extent, been due to the systemic implementation of nonisotopic affinity cytochemistry (AC), mainly immunocytochemistry (IC) and in situ hybridization (ISH). If we disregard, for example, the use of green fluorescent protein constructs allowing for the dynamic images of the nucleus, then, while extremely potent, the standard IC and ISH in situ usually suffer a drawback. The results obtained attest to the steady state of the cell. In other words, if we map a specific enzyme, we are unsure as to whether the mapped enzyme molecules perform their enzymatic job at the specific nuclear sites, or whether they are being recycled or present in a sort of storage pool etc. There are, with some reserve, exceptions to this rule such as the implementation of pulse-chase experiments in specific cases. An important exception are the mapping experiments of replications sites in which we can directly label through, e.g., incorporated bromodeoxyuridine (BrdU), during short incubation times, the active process of DNA synthesis itself (Gratzner 1982). Highly specific monoclonal antibodies to the brominated base are used for the detection of the replication signal. To a large extent, the same argument as for replication holds true for the mapping of transcription sites where the modified ribonucleosides incorporated into newly synthesized RNA are mapped. The same antibodies used for the detection of BrdU are used for the detection of incorporated bromouridine (BrU) in this latter case.

On a positive note, the standard affinity cytochemistry is capable of producing convenient results if we map a slow, rate-limiting step within the context of some cellular process. Most importantly, there is such a limiting step with respect to the synthesis and processing of pre-rRNA: the transcription time of pre-rRNA itself since this time of transcription is somewhere between 2.5 and 8 min in mammalian cells (Grummt 1978; Cavanaugh and Thomson 1985; Dundr et al. 2002). In addition, the "Christmas trees" are, with regard to the biogenesis of rRNA, the known nucleolar structures that remain in a steady state. There should be about 100 nascent pre-rRNA strands associated with the fully active ribosomal gene (e.g., Hadjiolov 1985). In this sense, if we map the transcription signal through the IC, an individual "Christmas tree" that likely contains thousands of incorporated modified ribonucleosides into nascent RNAs represents a convenient target par excellence. Most importantly, as far as I know, everybody accepts the existence of the "Christmas trees" in nucleoli of metabolically active animal and plant somatic cells.

However, a major problem encountered during in situ nucleolar studies has to be mentioned. The compactness of nucleoli sometimes makes it difficult to achieve the goals of cell biology studies, concerning the implementation of both nonisotopic IC and ISH approaches. The LM detection of incorporated BrU in newly synthesized nucleolar RNA may serve as a typical example (Koberna et al. 1999, 2000). In the case of the BrU detection, one fails to detect the incorporated brominated base in a standard way if the cells are fixed in formaldehyde, and this despite consecutive detergent treatment of cells (Wansink et al. 1993; Raška et al. 1995; Masson et al. 1996; Koberna et al. 1999, 2000; Fig. 4). However, RNA polymerase I "likes" the BrUTP (5-bromouridine 5'-triphosphate) and much of the incorporated BrU is present in the nucleolus, but the brominated bases are not detected within nucleoli by means of the IC. Other types of cell processing are necessary, such as methanol/acetone fixation, resulting in a looser nucleolar structure, which enables us to observe the BrU signal (Koberna et al. 2000; Staněk et al. 2001). At the same time, one has to keep in mind possible artifacts due to structural rearrangements of nucleoli and verify whether data obtained as a result of different fixation procedures generate compatible images within the framework of a given study. For instance, the detection of the movement of newly synthesized nucleolar RNA labeled with biotin provides, following either formaldehyde or methanol/acetone fixations, compatible results on the resolution of LM, while changes in the nucleolar structure are evidence of the two kinds of fixation on the ultrastructural level (Staněk et al. 2001).

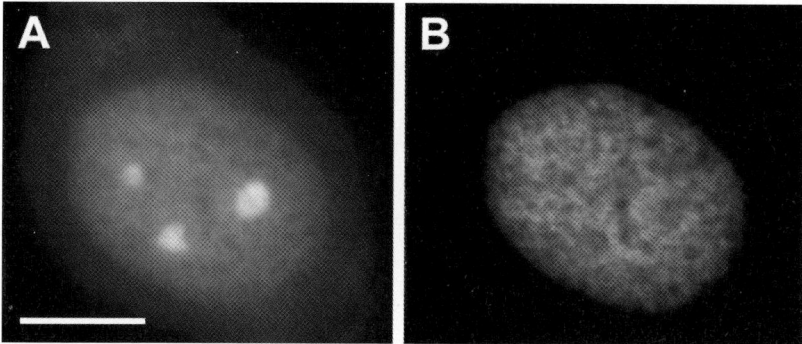

Fig. 4. LM mapping of transcription sites in HeLa cells. In this experiment, the permeabilized cells were incubated shortly with two different modified ribonucleosides (biotinylated and brominated uridine triphosphates) and fixed in formaldehyde. The incorporation of modified uridines was then revealed in two different colors by the IC. Note that in contrast to incorporated biotinylated uridine (**A**), no signal due to incorporated BrU (**B**) is seen in the nucleolus. The straightforward explanation is that while RNA polymerase II (and eventually RNA polymerase III) does incorporate brominated uridine into newly born RNA in the extranucleolar space, this is not the case of RNA polymerase I in nucleoli. However, this is a biased explanation. The BrU is namely incorporated into rRNA within nucleoli, but is not detected by the IC. *Bar*, 5 μm. Reproduced from Koberna et al. (1999) (Fig. 5f, g) by kind copyright permission of Springer-Verlag

One matter is worthy of emphasis in this respect. The LM nucleolar transcription BrU signal is revealed in formaldehyde-fixed cells if the treatment with α-amanitin is applied concomitantly (for discussion, see Koberna et al. 1999, 2000). This is apparently due to the rearrangement of the nucleolar structure ensuing from the drug treatment so that the brominated base can be detected by the standard IC. I believe that this is an important message for the reader in order to realize the possible dangers hidden behind the given experimental protocol.

In the nucleolar context, a comparison between LM and EM requires further comment. Pre-embedding AC EM results of thin-sectioned nucleoli can, to a large extent, be conceptually compared with parallel LM fluorescence results performed on whole cells (Humbel et al. 1998, Raška, unpubl. obs.). However, one has to be extremely careful about the impact of fine structure preservation in the pre-embedding EM experiments as the fine structure is always, either to a lesser or larger extent, rearranged (and extracted) with regard to the post-embedding approach (Raška et al. 1990, 1995; Melčák et al. 1996; Humbel et al. 1998; Gonzalez-Melendi et al. 2001; Fig. 5). Most importantly, the pre-embedding mapping results should always fit those of the parallel post-embedding EM approach (see below; Fig. 5). On the other hand, the pre-embedding EM experiments allow for important straightforward information with regard to such architectural nuclear elements, which are not prone to structural rearrangement and/or extractions.

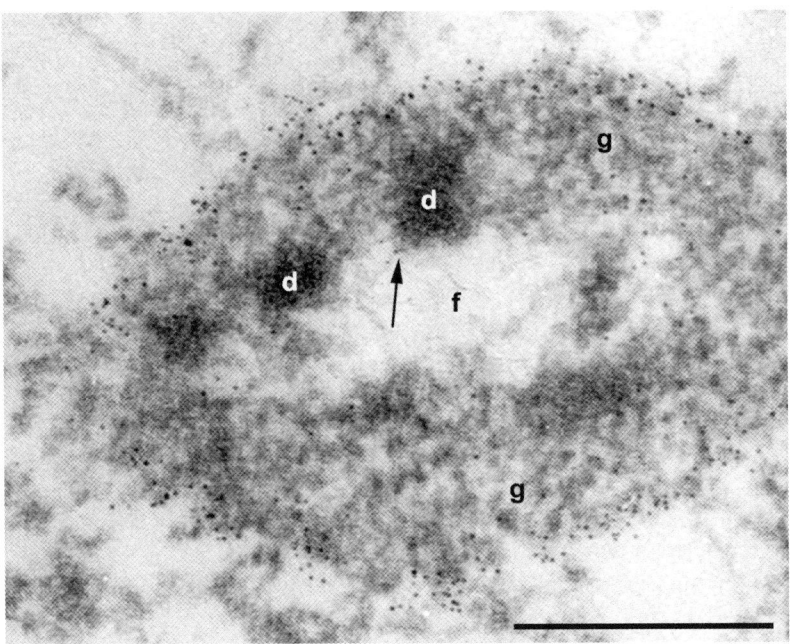

Fig. 5. Mapping of the ribosomal protein S1 in thin-sectioned rat hepatocyte nucleolus. The pre-embedding labeling was performed with the use of the monoclonal antibody to S1 protein (Hügle et al. 1985) and the secondary 5-nm gold adduct. Note that the ultrastructure is extensively changed, e.g., with regard to the nucleolus shown in Fig. 1. The nucleoplasm is largely rearranged and extracted. The nucleolus is partially preserved and one can distinguish the FCs, the DFCs and the GCs. Most gold particles only decorate the GCs at the circumference of the nucleolus in the section, and a few gold particles also decorate the edge of the DFCs within the extracted FC (*arrows*). This attests for a penetration problem of the primary antibody and/or secondary gold adducts into the DFCs and GCs. In contrast, if the S1 protein labeling is performed by the post-embedding procedure (Raška et al. 1990), the gold particles are seen within the DFCs (and the GCs), but not in the FCs. *f, d* and *g* correspond to fibrillar centers, dense fibrillar components and granular components, respectively. *Bar* 0.35 μm

The post-embedding AC EM approaches, which involve a standard preliminary fixation and in which we depict specific protein or nucleic acid markers exposed at the surface of sections, represent a more convenient approach (Raška et al. 1990, 1995; Melčák et al. 1996). Every microscopist knows that the convenient fixation necessarily also induces artifacts, but such artifacts prevent major artifacts taking place during further processing of cells and tissues. The number of accessible target structures on the surface of plastic sections or thawed cryosections is, of course, limited. This disadvantage is, to some extent, counterbalanced by the sectioning procedure which helps to expose the target macromolecules on the surface of sections and thus facilitates their detection. A typical example is the mapping LM experiments of the incorporated BrU in cells fixed in aldehydes (Fig. 6). While we have difficulties demonstrating the fluorescence nucleolar signal in experiments performed on whole aldehyde fixed cells (Fig. 4), a convenient fluorescence (as well as immunogold) nucleolar signal is seen in parallel experiments with aldehyde fixed and thin-sectioned embedded (or cryosectioned) cells (Koberna et al. 2002; Fig. 6). It should also be mentioned here that if the cells are fixed in a standard way in aldehyde, then the primary antibodies and secondary gold adducts do not even penetrate into thawed thin cryosections and the labeling is limited only to the structures exposed on the cryosection surface (Stierhof et al. 1986; Stierhof and Schwarz 1989; Griffiths 1993; Raška, unpubl. obs.). It should also be mentioned here that the fine structure of nucleoli seen after standard fixation is entirely compatible with that of rapidly frozen (= cryofixed) and freeze substituted cells devoid of artifacts (e.g., von Schack et al. 1991).

Importantly, the nucleolar FCs frequently appear less compact and more prone to extraction with respect to the DFCs (Fig. 5). The efficiency of AC detection of the target macromolecules may generally differ between the FCs and the DFCs (Raška et al. 1990, 1995; Raška, unpubl. res.).

Last, but not least, I emphasize that the EM, but not always LM, allows for the correct interpretation of the AC results for structures the dimension of which is on the edge of the resolution limit of LM (up to 300 nm in the case of the FC dimensions in HeLa cells). In a case involving the mapping of nucleolar transcription in HeLa cells, not the restored fluorescence images, but the EM images show the genuine distribution of the transcription signal (Malínský et al. 2002). In this sense, the LM investigation of nucleoli should be accompanied by the parallel EM approach (Mais and Scheer 2001; Dundr et al. 2002).

4
Mapping of Active Ribosomal Genes

Keeping in mind nucleolar compactness, I have to emphasize that there is no direct way to exclusively label the active ribosomal genes in situ. I shall now enumerate various nonisotopic approaches, the results of which are used as arguments for the localization of active genes in situ, and comment on the respective relevance of the results obtained. I shall discuss the results with regard to the FC and DFC only, avoiding the GC. I also must point out that, in contrast to the spreading technique, the available evidence attests for compacted and contorted "Christmas trees" in the thin-sectioned nucleoli (Trendelenburg 1974; Hadjiolov 1985; Tröster et al. 1985; Raška et al. 1995; Melčák et al. 1996; Scheer et al. 1997; Gonzalez-Melendi et al. 2001; Koberna et al. 2002). Due to a number of associated proteins and nonribosomal RNPs, the actual extent of compactness of the "Christmas trees" may, of course, differ among different species or different types of cells.

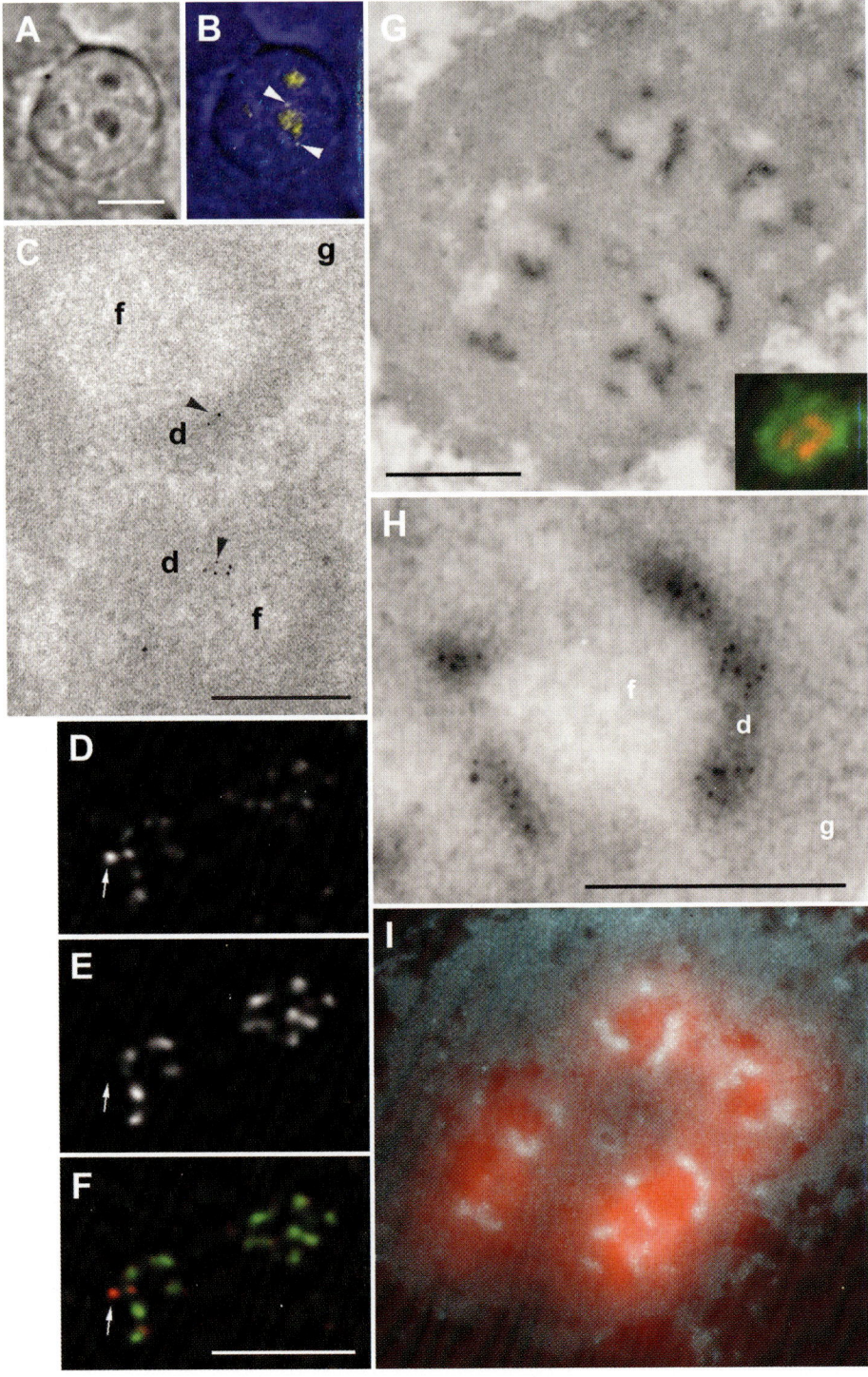

4.1
Electron Microscopy and Cytochemistry Approaches

If one accepts the existence of "Christmas trees" within the nucleolus of metabolically active somatic cells of higher eukaryotes, then a straightforward common sense argument follows. The "Christmas trees" generate such a high local concentration of RNA, respectively of RNPs that they have to be seen by direct EM in the nucleolar FCs of conventionally stained ultrathin sections of the nucleolus. On the other hand, if present in the DFCs, it is hard to see the "Christmas trees" due to the electron-dense nature of this nucleolar subcompartment.

The direct visualization of the "Christmas trees" in situ was provided by Scheer et al. (1997) in the nucleoli of oocytes, and particularly in isolated oocyte nuclei, of grasshoppers. The oocyte nucleolus has a special morphology and the "Christmas trees" are clearly seen within the so-called nucleolar pocket. The electron lucent nucleolar pockets possibly correspond to the FCs (Scheer et al. 1997), but in my opinion, their ultrastructure differs from the

----

Fig. 6. Localization of ribosomal genes and nucleolar transcription in HeLa cells. **A, B** Wide field LM ISH mapping of ribosomal genes. Fluorescence mapping of ribosomal genes (**B**, in *yellow*) and phase contrast image (**A**; *blue* in **B**) are shown. Ribosomal genes are located in a number of fluorescent foci (in general from 10–40 foci per nucleus). Most of the foci are situated in the nucleolus. Two foci are apparently situated in the perinucleolar chromatin (*arrowheads*). **C** EM mapping of ribosomal genes in a thin-sectioned nucleolus. The signal is in general low, but highly specific. In this particular section, the signal that apparently consists of just two hits (the two clusters of gold particles are designated by *arrowheads*), is entirely located in the DFCs. The message from this image is that rDNA is also present in the DFCs. **D–F** Light microscopic image of ribosomal genes (**D**; red in **F**) with respect to the position of nucleolar transcription sites (**E**; green in **F**). The merged image is in **F**. One confocal section after image restoration is shown. Most of the rDNA foci exhibit the transcription activity. An rDNA focus without transcription signal is indicated by the *arrow*. **G–I** Concomitant LM and EM mapping of transcription signal in a thin section of permeabilized cell. **G** Nucleolar transcription signal consists of numerous clusters of gold particles accumulated in the DFCs. The clusters are indicative of the presence of "Christmas trees" in the DFCs and the DFC/FC borders. The stereological analysis of the clustered transcription signal in **G** further suggests that the "Christmas trees" are contorted in space and exhibit a DNA compaction ratio in the order of 4–5.5. **H** is a detailed part of **G**. The *colored insert* in **G** represents the wide field LM picture of the same area depicting newly synthesized RNA (*red*) and nucleic acids counterstained with YOYO (*green*). **I** shows the merging of an inverse EM image from **G** in cyan, and a transcription LM signal from the upper insert in *red*. Note the overlap of the LM and EM transcription signals. In fact, the LM signal is present not only over the DFCs, but also over the FCs and the proximal part of the GCs. LM analysis does not permit the FCs to be distinguished from the DFCs and LM transcription signal is thus mapped to both FC and DFC (as well as the proximal part of GCs). Importantly, we have shown (Malínský et al. 2002) that after the deconvolution of the LM signal shown in **I**, one is tempted to interpret the transcription signal in a biased way. The deconvoluted signal is primarily seen in the FCs, not in the DFCs (Malínský et al. 2002). *Bars* in **A, B** 10 μm, in **C, H** 0.2 μm, in **D, E, F** 5 μm, in **G, I** 0.5 μm. Reproduced from Koberna et al. (2002) (Fig. 1B, C, D, Fig. 2C, D, E, Fig. 4A, B, C) by kind copyright permission of the Rockefeller University Press

FCs encountered in animal somatic cells. I have three comments with regard to this work. At high magnification (Fig. 2A in Scheer et al. 1997), the DFCs may also contain myriads of the "Christmas trees". There are many thousands of active ribosomal genes in the form of the "Christmas trees" in the grasshopper oocyte nucleus (Scheer et al. 1997) and I can distinguish a high number of structures (of the size and form of the "Christmas trees" observed in the electron lucent nucleolar pocket) in the DFCs, the electron density of which is higher than that of the DFCs in their vicinity. Importantly, if the "Christmas trees" are not seen within the nucleolar pocket next to the DFC, I would never be aware of these "denser" structures within the DFCs. Second, the nonisotopic incorporation of BrU on the EM level is not performed. As the authors use isolated nuclei (see Mapping of the Nucleolar Transcription Signal), the BrU incorporation approach represents the approach of choice. Finally, the authors conclude that the possibility to see the "Christmas trees" much better in the isolated nuclei in comparison to whole cells is due to the slow fixation of dissected ovarioles that probably allows for the structural rearrangements. This is a valid argument. However, an equally valid argument can be made that the isolation of nuclei can lead to the accentuated dissociations of the "Christmas trees" from the DFC into the nucleolar pockets. However, this paper is of primary importance, it brings the first genuine in situ description of the "Christmas trees", and leaves the reader without any ambiguity with regard to the existence of highly compacted "Christmas trees" that do not exhibit a laterally polarized structure in situ.

There are a number of publications devoted to the cytochemistry of the nucleolus (e.g. Marinozzi and Bernhard 1963; Bernhard 1969; Busch and Smetana 1970; Smetana and Busch 1974; Fakan 1986; Derenzini et al. 1990; Testillano et al. 1991; Risueno and Testillano 1994). DNA, RNA or various proteins have been mapped in these studies. While RNA is hardly detectable in the FCs, the DFCs are rich in RNA and DNA is mapped by different groups either to the FCs or the DFCs or both nucleolar subcompartments. The most commonly implemented technique is likely the Feulgen-like osmium amine staining for DNA (Cogliati and Gautier 1973; Derenzini et al. 1990). Nobody argues against the presence of DNA in the FCs. The relevant message following from these studies by many research groups is that DNA is mapped to the DFCs as well. In addition, Biggiogera et al. (2001) have combined the immunocytochemistry of fibrillarin, a specific protein marker of the DFC (Christensen et al. 1977; Ochs et al. 1985; Reimer et al. 1987; Raška et al. 1989; Puvion-Dutilleul et al. 1991b), with energy filtering transmission electron microscopy of DNA. The authors describe a "DNA cloud" consisting of an inner core of DNA fibres, the FC, and a periphery made of extremely thin fibrils overlapping with the fibrillarin-positive areas, the DFC (Biggiogera et al. 2001).

Silver staining techniques represent important techniques used to identify the NORs in metaphase chromosomes (e.g. Goodpasture and Bloom 1975; Hernadez-Verdun et al. 1978; Schwarzacher et al. 1978; Howell 1982; Spector et al. 1884; Derenzini et al. 1990). This approach is of importance both in basic

research and human pathology and the positive chromosome silver precipitates testify to the presence, in the corresponding NOR, of active ribosomal
genes during previous interphase (Sirri et al. 2000; Sullivan et al. 2001). However, if the technique is used in interphase cells on the EM level, the silver
precipitates can be seen both in the FCs and (mainly in) the DFCs (e.g. Moreno
et al. 1985; Biggiogera et al. 1989; Derenzini et al. 1990; Wei et al. 2003).

4.2
Immunocytochemical Mapping of Nucleolar Chromatin Structures

The results obtained by the IC of chromatin structures such as double stranded
and/or single stranded DNA or other chromatin components can be summa-
rized as follows: either the FCs or the DFCs or both nucleolar subcompartments contain chromatin structures (Thiry and Müller 1989; Raška et al. 1990,
1995; Thiry 1991; Thiry and Goessens 1991; Mosgöller et al. 1993; Raška and
Dundr 1993; Thiry et al. 1993; Vandelaer et al. 1993; Risueno and Testillano
1994; Testillano et al. 1994; Gonzalez-Melendi et al. 1998). These studies also
include the implementation of the nick-translation system or the use of the
terminal deoxynucleotidyl transferase reaction that provide a convenient
amplification of the signal due to DNA. In addition, the newly replicated DNA
is also found in the DFCs (Koberna and Raška, unpubl. res.).

All these studies have provided a highly specific labeling of chromatin structures. However, at least in mammalian cells, ribosomal genes represent a very
minor fraction of the total intranucleolar DNA (Bacchellerie et al. 1977) and
most of the signal observed is irrelevant with regard to the active ribosomal
genes. Similarly to the statement of the previous section, everybody agrees that
chromatin structures are present in the FCs. On the other hand, the majority
of groups working in the nucleolar field have been able to detect DNA in the
DFCs as well. I am of the opinion that this is the relevant message of this
subsection (for a more detailed discussion, see also Raška et al. 1995).

4.3
In Situ Hybridization Mapping of rDNA

In contrast to the previous two subsections, the ISH detection of rDNA already
provides the specific approach. Numerous groups have mapped rDNA into the
FCs and/or the DFC (e.g. Thiry and Thiry-Blaise 1989; Puvion-Dutilleul et al.
1991a; Wachtler et al. 1992; Jimenez-Garcia et al. 1993; Raška et al. 1995;
Gonzalez-Melendi et al. 2001; Koberna et al. 2002; Fig. 6). Unfortunately, these
mapping experiments alone cannot provide an answer as to whether the
mapped gene is active or not.

It is also worthwhile here to comment on the quantitative aspect of ISH
mapping in the post-embedding EM approach. In cultured mammalian cells,

for instance, there should be at most only a few active genes per complex consisting of one FC together with the surrounding DFC (Koberna et al. 2002). The chance that a sufficiently long sequence of rDNA is exposed at the surface of thin sections, and is consecutively detected through ISH is thus very low (Fig. 6). Therefore, the corresponding signal in a single thin-sectioned nucleolus has to be extremely low anyway and only a fraction of the FC/DFC complexes should, in a single nucleolar section, exhibit a signal on the EM level due to (active) ribosomal genes.

We attempted to map concomitantly rDNA and the transcription signal. While we were able to obtain the standard results on the level of LM of whole cells, we were, for the time being, unsuccessful on the ultrastructural level with the post-embedding approach on thin-sectioned nucleoli. Fluorescence results testify to the fact that in HeLa cells ribosomal genes cluster 10–40 fluorescence foci per nucleus and that most foci are transcriptionally active (Koberna et al. 2002; Fig. 6).

4.4
In Situ Hybridization Mapping of rRNA

This category of results is, of course, of utmost importance, but again, suffers from a drawback. In standard mammalian cell lines for instance, there are many tens or even hundreds of ribosomal subunits exported each second (Görlich and Mattai 1996) and the straightforward mapping of rRNA sequences reflects their steady state distribution. Therefore, the relevance of the rRNA mappings to the active gene mappings has a limited impact only. The rRNA mapping results attest to the presence of transcripts in the DFC together with the DFC/FC border zone, while only a very minor or no signal at all is seen in the interior FCs (Puvion-Dutilleul et al. 1991a; Wachtler et al. 1992; Olmedilla et al. 1993; Raška and Dundr 1993; Raška et al. 1995; Thompson et al. 1997). In addition, the 5S rRNA that is synthesized outside the nucleolar body by the RNA polymerase III, but appears in the nucleolus later, is mapped to the DFC, but not to the FC (Raška et al. 1995).

One may always counter with the argument that in animal somatic cells (note that there is no controversy with regard to plant somatic cells) rRNA is synthesized in the FCs, but is released after completion of transcription, and is rapidly translocated into the DFCs where it is accumulated. However, where is then, within the interior of the FCs, the label due to the "Christmas trees" themselves? In addition, the intense transcription signal consisting of clustered gold particles in the DFCs (Fig. 6) should point to a special transport mechanism for released RNAs.

Within the context of rRNA or transcription signal (see below) mappings, one can also implement experiments in which the rRNA biogenesis is altered through drugs (e.g. actinomycin D, dichloro-ribofuranosylbenzimidazole (DRB) or cordycepin) that suppress transcription or elongation of transcripts

(Busch and Smetana 1970; Stockert et al. 1970; Noaillac-Depeyre et al. 1989; Raška et al. 1990; Haaf and Ward 1996). It is well known, however that the use of such drugs quickly leads to a rapid and spectacular rearrangement of the nucleolar structure and one has to be careful about the interpretation of the data obtained. Even the use of α-amanitin, which primarily does not affect the RNA polymerase I-mediated transcription, leads to the structural rearrangement of nucleoli (Sinclair and Brasch 1978; Haaf and Ward 1996). Importantly, however, in this report I will not look at the modulation (including re-initiation) of the rRNA synthesis, due either to the drug treatment and the cell entering into interphase after mitosis, nor will I look at the fate of ribosomal genes or transcripts after the modulation of transcription. In this report, I will only look at the steady state of nucleoli seen under physiological conditions, and everybody agrees that this state corresponds to the presence of "Christmas trees" within the nucleolus.

Within the context of this subsection, an important approach is represented by the mapping of transcribed ribosomal sequences, which are cleaved away during the processing of pre-rRNA, i.e., external and internal transcribed spacers (Fischer et al. 1991; Puvion-Dutilleul et al. 1991a, 1992, 1997; Mena et al. 1994; Beven et al. 1996; Lazdins et al. 1997; Dundr and Olson 1998; Staněk et al. 2001). The cleavage consists of the cascade of steps, separated, depending on the cell type, more or less in time. It is worth mentioning that the components involved in the very early cleavage that involves the 18S rRNA associate with the nascent pre-rRNA (Scheer and Benavente 1990; Mougey et al. 1993). Importantly, the 5′externally transcribed spacer RNA sequences are mapped only to the DFCs and the DFC/FC border of mammalian nucleoli (Puvion-Dutilleul et al. 1991a, 1997). In yeast, the 5′ part of the externally transcribed spacer 1, which is found in the very early pre-rRNA precursors, is mapped by ISH in the DFCs, but not in the FCs (Trumtel et al. 2000).

Sequential cleavage involving a time lapse of several minutes was used by Lazdins et al. (1997). In this study, importantly, the authors also determined the half-life of respective cleaved products. Using two different fluorochromes, they concomitantly mapped the distribution of two cleaved sequences in mouse cells on the LM level, one sequence at the 5′ end of the transcript being cleaved very early, the second at the 3′end several minutes later. The 3′ end cleaved product is synthesized as the latest part of the pre-rRNA. The two cleaved sequences have a half-life of only 2 min. The ISH mapping results of the 5′ end of the pre-rRNA identify the nascent transcripts and/or first steps of the processing, while the mappings of the 3′ end of pre-rRNA identify mature transcripts and/or their processing steps at the 3′ end. The conclusion of the authors designates the DFCs, and not the FCs, as the site of synthesis and/or processing pre-rRNA (Lazdins et al. 1997). This work was recently confirmed through an alternative approach by Staněk et al. (2001) who mapped the 5′ externally transcribed spacer together with the transcription signal into the DFCs, but not the FCs, of HeLa cell nucleoli. In addition, experiments performed in cells that are administered modified nucleosides in vivo (Staněk et al. 2001) demonstrate the vectorial movement of released rRNA transcripts from the DFC towards the GC (as well as the appearance of signal in the cytoplasm later on) according to the same kinetics as described in the pioneering autoradiographic work of Granboulan and Granboulan (1965). The movement of transcripts reported in Staněk et al. (2001) is in agreement with the results of Cmarco et al. (2000), but are partially at variance with those of Thiry et al. (2000). In this last communication, the lipofectin (FuGene) approach is used for the delivery of bromouridine triphosphate (BrUTP) into the cells. It requires 15 min of incubation under conditions that allow incorporated BrU detection so that the actual time of BrU incorporation remains elusive; in addition, the treatment with α-amanitin is used. The authors (Thiry et al. 2000) consider the FCs to be the primary sites of rDNA transcription, but the transcription signal within the FCs is, in the pictures shown, extremely low.

4.5
Immunocytochemical and In Situ Hybridization Mapping of Nucleolar Proteins and Ribonucleoproteins

Similar to what has been said in the previous subsection, the mapping of ribosomal and nonribosomal proteins and nucleolar RNPs primarily provides the view of the steady state. It is difficult to say that a given macromolecule that is known to be associated with the transcription process, directly participates in the active process of transcription. Nevertheless, there are a few helpful markers and the best known, with respect to the mapping of active genes, is the protein fibrillarin.

There is no standard marker of the FCs. In contrast, fibrillarin is involved in the early processing of the pre-rRNA, is a component of, e.g., U3 snoRNP, and is mapped to the DFCs, but not to the interior of the FCs (Ochs et al. 1985; Reimer et al. 1987; Raška et al. 1989; Kass and Sollner-Webb 1990; Kass et al. 1990; Savino and Gerbi 1990; Scheer and Benavente 1990; Puvion-Dutilleul et al. 1991b; Mougey et al. 1993; Borovjagin and Gerbi 2001). These results apparently limit the functional involvement of fibrillarin to the exterior of the FCs. Importantly, it has been shown that fibrillarin maps at sites of pre-rRNA synthesis (Garcia-Blanco et al. 1995; Cmarko et al. 2000) and that RNA polymerase I associates with the processing machinery (Fath et al. 2000).

RNA polymerase I, the key molecule with regard to the transcription, is mapped preferentially to the FCs, but also to the DFCs (Scheer and Rose 1984; Scheer and Raška 1987; Raška et al. 1989, 1995; Fig. 3). Interestingly enough, this enzyme is mapped preferentially to the DFCs if some other source of antibodies is used (Cmarko et al. 2000). With respect to HeLa cells (cell cycle of about 1 day), in more rapidly proliferating cells (cell cycle of up to 10 h), the RNA polymerase I signal in the DFCs is relatively higher (Raška et al. 1995). In yeast, RNA polymerase I is mapped into the DFCs, but not to the FCs (Léger-Silvestre et al. 1999).

The FCs apparently play an architectural role in the arrangement of ribosomal genes in animal somatic cells. In this respect, it has been shown that RNA polymerase I sits on the nontranscribed spacer in the spread ribosomal gene units (Chooi and Leiby 1983). Thus, the RNA polymerase I signal in the FCs may be partly due to such nontranscribing RNA polymerase I molecules that, however, sit on the rDNA.

In addition, another category of proteins has to be mentioned here, namely ribosomal proteins. These proteins, or at least some of them, have been shown to bind the nascent transcripts (Chooi and Leiby 1981) and they map to the DFCs (as well as the GCs) and not to the FCs (Raška et al. 1990, 1992, 1995).

A high number of nonribosomal protein factors and snoRNAs, (several snoRNAs offer the possibility to map their trimethylguanosine epitope) that are involved in the synthesis and/or processing of pre-rRNA have been mapped in the last 20 years (e.g., Gas et al. 1985; Ochs et al. 1985; Reimer et al. 1987, 1988; Schmidt-Zachmann et al. 1987; Biggiogera et al. 1989, 1990; Raška et al. 1989, 1992, 1995; Clark et al. 1990; Fischer et al. 1991; Puvion-Dutilleul et al. 1991b; Meier and Blobel 1992, 1994; Rendón et al. 1992; Matera et al. 1994; Roussel et al. 1993; Jacobson et al. 1995; Beven et al. 1996; Jordan et al. 1996; Kill 1996; Zatsepina et al. 1997; Mosgöller et al. 1998; Bunney et al. 2000; Chen and Huang 2001; Tuteja et al. 2001; Iben et al. 2002). This allows for a detailed view of the nucleolus with, however, only a supportive indication with regard to the localization of active genes. These molecules are usually mapped both into the FC and the DFC, but in frequent cases, their incidence is much higher in the DFCs, together with the DFC/FC border, than in the interior of the FCs.

4.6
Mapping of the Nucleolar Transcription Signal

I am of the opinion that the nonisotopic mapping of the transcription signal (Dundr and Raška 1993; Jackson et al. 1993; Schöfer et al. 1993; Wansink et al. 1993; Hozák et al. 1994; Figs. 1, 6, 7) represents the method of choice with regard to the localization of active ribosomal genes in animal and plant somatic cells. This approach does not identify active ribosomal genes directly, but only indirectly through the intensive signal in nascent pre-rRNAs. In comparison to autoradiography, this approach is rapid and provides much higher resolution.

The drawback of this approach (as well as that of autoradiography) is that even during a short incubation period some transcripts are finished, and that we may, in fact, map the accumulations of released transcripts and not the nascent ones. My counter-argument is that, regardless of what the number of released transcripts is, at least a sizeable part of the transcription signal is to be associated with the genuine "Christmas trees". I consider this possible drawback that reminds me the proverb "you cannot see the forest for the

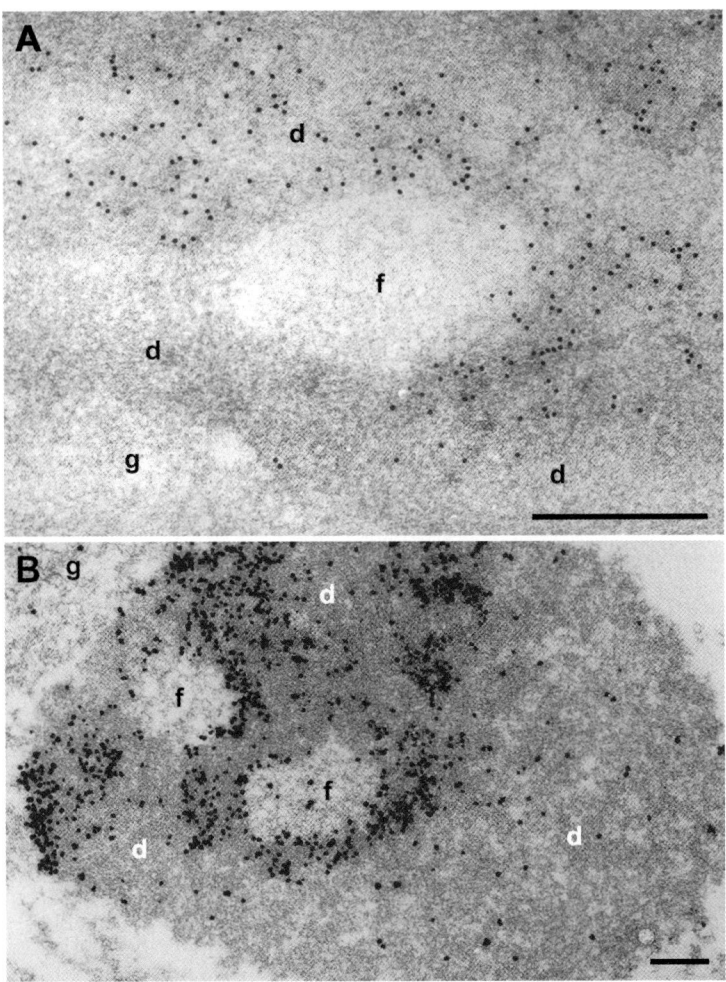

Fig. 7. Localization of nucleolar transcription sites in permeabilized onion root tip protoplasts. A The post-embedding localization of RNA synthetic sites (10-nm gold particles). Most gold particles are present in the DFCs. B The pre-embedding IC of the transcription signal (silver intensified 1 nm gold adduct). Most silver particles that are frequently clustered are present in the DFCs and the labeling pattern is compatible with A, but is much higher. Nevertheless, the nucleolar ultrastructure suffered from the pre-embedding processing of protoplasts. Note that the subdomains of the DFCs in A and B that exhibit gold (A) or silver (B) particles are more electron-dense than the remaining part of the DFCs. *Bars* 0.3 μm. After Melčák et al. (1996) (Fig. 1E, Fig. 3D) by kind copyright permission of Elsevier

trees",[3] as valid, but not entirely applicable with regard to the results obtained (Figs. 1, 6, 7).

This entire report thus basically stands only on two matters: that one accepts the existence of more or less compacted "Christmas trees" in the nucleolus, and that, through the transcription signal, one can also identify the active genes, not just released transcripts. As already mentioned, the available evidence points to the existence of such "Christmas trees" in nucleoli of metabolically active somatic cells, so I do not consider the first matter to represent a real constraint. The second matter represents the real constraint which, in my opinion, is not that strong. However, we shall see what the future holds. At present, I have to leave the decision of whether the second constraint is acceptable up to the reader.

Several approaches are used for the visualization of transcription sites. In the first, BrUTP, or some other modified ribonucleoside triphosphates, are administered to the cells which have been permeabilized (e.g., streptolysin O or detergent are used). After a short incubation period, usually up to 5 min, the BrU incorporated into RNA is revealed by the standard IC, both at LM and EM (Fig. 6). I have to comment on this approach as it has several peculiarities. The initiation of rDNA transcription is lowered or even blocked, but the elongation somehow proceeds, and the sizeable incorporation rate lasts for many minutes (Jackson et al. 1993, 1998; Wansink et al. 1993; Staněk et al. 2000). In addition, the processing of pre-rRNA, at least to some extent, takes place and the transcripts remain at the site of transcription (Staněk et al. 2000; Koberna et al. 2002). The ultrastructure of the nucleoplasm, and particularly of the cytoplasm, suffers from the treatment. However, even though it also has to necessarily suffer from some extractions, the nucleolar ultrastructure is well preserved (Dundr and Raška 1993; Raška et al. 1995; Staněk et al. 2001; Koberna et al. 1999, 2002; Fig. 6) and at the very least, it remains in the same good condition as isolated nucleoli. Importantly, the intense transcription immunogold signal in the DFCs together with the DFC/FC border consists of clusters of gold particles and is confined to the subdomains of the DFCs only. This testifies to the functional compartmentation of the DFCs (e.g. Dundr and Raška 1993; Melčák et al. 1996; Lazdins et al. 1997; Mosgöller et al. 1998; de Carcer and Medina 1999; Staněk et al. 2001). The remaining part of the DFCs that is not engaged in transcription, apparently serves for the later steps in the processing of rRNA, for the assembly of the maturation/transcription complexes and/or as a reservoir of necessary rRNA processing and modifying factors (Lazdins et al. 1997; Mosgöller et al. 1998; Staněk et al. 2001).

In a modified protocol for the visualization of transcription sites, the cells are not permeabilized, but thicker (e.g., vibratome) sections from the tissue are cut so that the cells are opened mechanically. The sections are then incubated with modified ribonucleoside triphosphates and processed, either for

---

[3]I do not want to mislead the reader, such trees have, of course, nothing in common with the "Christmas trees"

the pre-embedding or post-embedding IC, in a similar way as above. The intense clustered transcription signal is confined to the DFCs together with the DFC/FC border (Gonzalez-Melendi et al. 2001).

The next approaches explore cells that have incorporated modified ribonu-cleosides in vivo. In one approach, used by the group of Peter Cook, not the BrUTP, but BrU is administered to the cells and following a short incubation period its incorporation sites are revealed by the standard IC (e.g., Iborra et al. 1998; Jackson et al. 1998).

In another approach, the modified ribonucleoside triphosphates, such as BrUTP, are introduced in the living cells through, for example, the transfection kits such as FuGene, and after a short incubation period, the incorporated markers are revealed by the standard IC. With regard to the FuGene kit, we have noticed (Koberna et al. 1999, 2000) that the use of the hypotonic buffer of the FuGene kit by itself is, with regard to the use of the whole FuGene kit, more convenient and less deleterious for the delivery of modified nucleoside triphosphates into the cells. We have, therefore, standardized the protocol in which the cells or small pieces of tissues are shortly exposed to a cold hypo-tonic buffer that contains modified nucleosides. No incorporation of BrU takes place during the hypotonic shift. In the normal medium at the standard tem-perature of incubation, we then begin to observe, in the cultured cells, the sufficiently intense fluorescence transcription signal after 2 or 3 min and keep the incubation times up to 5 min. In practical terms this limits the actual incorporation time of modified ribonucleosides to 2 or 3 min only. Impor-tantly, the ultrastructure of nucleoli after the 5-min incubation period reverts back to normal (Koberna et al. 1999, 2002; Staněk et al. 2001; Fig. 1). The intense nucleolar signal is seen in the DFCs and the DFC/FC border. This being said, how can this pattern of transcription be reconciled (Fig. 1) with that seen in permeabilized cells (Fig. 6)? At first sight, one might be struck by the differ-ences. However, the highly positive and most relevant message is that the intense signal is seen in the DFCs together with the DFC/FC border in both cases. Nevertheless, why is the signal in permeabilized cells higher and clus-tered? First, in the system of permeabilized cells, the quantity of available modified ribonucleosides is, in contrast to the latter approach, not limited and the detergent treatment, as already extensively discussed in this report with regard to the compactness of nucleoli, necessarily helps to expose the target epitopes and thus improves the efficiency of the IC. Second, the cells that have incorporated modified ribonucleosides in vivo exhibit similar features as non-treated cells with regard to newly synthesized nucleolar RNA (Staněk et al. 2001; Koberna et al. 2002), and the results obtained are in agreement with high resolution autoradiography documenting the kinetics of movement of newly synthesized RNAs (Granboulan and Granboulan 1965). Indeed, the released transcripts move away from the site of transcription (Staněk et al. 2001; Koberna et al. 2002). In permeabilized cells, the labeled RNAs remain at the site of transcription (Staněk et al. 2000; Koberna et al. 2002) and give rise to the highly localized signal in the form of clusters of gold particles.

The next approach explores the use of microinjections of nucleoside triphosphates (Wansink et al. 1993; Masson et al. 1996; Cmarko et al. 1999, 2000). This is an elegant approach in which the analysis of results is limited to the microinjected cells only.

Through all these approaches, the majority of groups in the nucleolar research field have seen the transcription signal in the DFC together with the DFC/FC border of both animal and plant somatic cells (Dundr and Raška 1993; Schöfer et al. 1993; Hozák et al. 1994; Raška et al. 1995; Melčák et al. 1996; Thomson et al. 1997; Mosgöller et al. 1998, 2001; Cook 1999; de Cárcer and Medina 1999; Cmarko et al. 2000; Gonzalez-Melendi et al. 2000; Figs. 1, 6, 7). Either no signal at all or a very minor one is seen in the interior of the FC. Importantly, Gonzales-Melendi et al. (2001) could unambiguously describe the presence of compacted "Christmas trees" in the DFCs of plant somatic cells. Recently, we have been able to obtain an intense transcription signal in the DFCs together with the DFC/FC border in the nucleoli of HeLa cells. The signal consists entirely of clusters of gold particles and is thus indicative of the presence of the "Christmas trees" (Koberna et al. 2002; Malínský et al. 2002).

There is a difference in the arrangement of active genes in the DFCs between animal and plant somatic cells. In the plant cells, the number of ribosomal genes is usually much higher (frequently many thousands of genes per nucleus as compared to hundreds of genes encountered in mammalian somatic cells; Long and Dawid 1980; Hadjiolov 1985) and the DFCs occupy most of the nucleolar area in section. The active genes in plant nucleoli map not only to the vicinity of the FCs, but are often found separated in space from the FCs (Melčák et al. 1996; Gonzalez-Melendi et al. 2001). In contrast, (the majority of) active ribosomal genes, as depicted through the transcription signal, are apparently more associated in space with the FCs in animal somatic cells (Dundr and Raška 1993; Schöfer et al. 1993; Hozák et al. 1994; Raška et al. 1995; Cmarko et al. 1999, 2000; Cook 1999) and I expect an architectural role of the FCs with regard to the arrangement of active ribosomal genes. In cultured mammalian cells, the individual NORs typically unravel and necessarily give rise to several complexes consisting of the FC together with the surrounding rim of the DFCs. In these cells, the DFCs form a sort of reticulum and a possibility exists that the (active and/or) inactive ribosomal gene(s) generate physical linkage(s) between the different FC/DFC complexes (Junera et al. 1995).

In order to complete the list, I need to mention the data in which the transcription sites are mapped into the FCs. In an alternative approach, Thiry et al. (2000) have put the FuGene reagent and the modified ribonucleoside triphosphates into the normal medium and observed, after a 15-min delay, the initial transcription signal in the FCs (see also the subsection on rRNA mappings). Similarly, Cheutin et al. (2002) have concluded that the active ribosomal genes are in the interior of the FCs. The authors analyzed the transcription signal in isolated nucleoli from Ehrlich ascites tumor cells and using these results, they perform a highly demanding tomographic reconstruction of active ribosomal genes based entirely on the RNA polymerase I pre-embedding mappings in a different cell line. I have a concern with regard to both approaches. First, the transcription signal seen in the FCs of isolated nucleoli is extremely low in the documented micrographs. In this respect, I personally find their previous autoradiographic results more convincing (Thiry et al. 1985). Another concern is the impact of the tomography in Cheutin et al. (2002). I believe this demanding work is performed on the basis of an inconvenient and difficult premise: the prototype micrograph of the RNA polymerase I mapping (Fig. 6 in Cheutin et al. 2002) is unacceptable for the strong and far-reaching claims of the authors.

Leaving the world of somatic cells, Mais and Scheer (2001) have mapped the transcription signal into the FCs in *Xenopus laevis* oocyte nucleoli at maturation stages that are highly active in the rDNA transcription. These oocytes contain a high number of active extrachromosomal rDNA copies confined to extrachromosomal nucleoli. My comment is that the authors do not perform the EM mapping of transcription sites. However, they are able to differentiate the FCs from the DFCs on the LM level.

On a positive note, the implementation of the nonisotopic labeling of newly synthesized rRNA in cells that have incorporated modified ribonucleosides in vivo, testifies to the vectorial movement of labeled transcripts from the DFCs to the GCs and then to the cytoplasm, respectively from the FCs to DFCs, then from the DFCs to the GCs and finally to the cytoplasm (Cmarko et al. 2000; Thiry et al. 2000; Staněk et al. 2001). This vectorial movement, reflecting the sequential steps in the processing of pre-rRNA and maturation of ribosomal particles is in agreement with the molecular biology and biochemical data (Fatica and Tollervey 2002).

5
Conclusions

In this study, I have presented my view on the organization of highly active ribosomal genes within the nucleolar architecture in order to address the long-standing controversy whether, in animal and plant somatic cells, such transcribing ribosomal genes are found in the DFCs (together with the DFC/FC border) or the FCs (together with the FC/DFC border). In this respect, I consider the nonisotopic visualization of the transcription signal as the method of choice. Of course, I remain concerned that my opinion is based on two important matters. First, that active ribosomal genes, or at least their sizeable part, are highly active, i.e., their transcription activity is compatible with the in situ existence of the "Christmas trees". As far as I know everybody agrees on this matter. The second matter represents a real constraint, namely that the mapping of transcription signal, due to a short pulse of modified ribonucleotide, also allows for the detection of such "Christmas trees".

Based on the results of a majority of groups working in the nucleolar field, I hold that highly active ribosomal genes are (part of) the DFCs. In plant somatic cells, there is a consensus that the DFCs together with the DFC/FC border are the nucleolar substructures to which the "Christmas trees" are confined. With regard to the animal somatic cells, my major arguments for the presence of "Christmas trees" in the DFCs, together with the DFC/FC border, are: (1) rDNA is present in the DFCs (Fig. 6C); (2) In cells that have incorporated modified ribonucleosides in vivo, the intense immunogold signal is seen in the DFCs together with the DFC/FC border (Fig. 1); (3) In permeabilized cells, the intense and clustered transcription signal is seen in the DFCs together with the DFC/FC border and is indicative of the presence of the "Christmas trees" (Fig. 6G–I); (4) There is a consensus that early pre-rRNA processing steps take place in the DFCs and that fibrillarin is mapped to the DFCs and the DFC/FC border, but not to the interior of the FCs, and it has been shown that some processing steps take place already on nascent transcripts in *Dictiostelium* and yeast (Grainger and Maizels 1980; Beyer, pers. comm.); (5) The mapped transcribed spacers, indicative of transcription sites and/or just finished transcripts, are mapped to the DFCs together with the DFC/FC border (Puvion-Duttileul et al. 1991a, 1997; Shaw et al. 1995; Beven et al. 1996; Lazdins et al. 1997; Staněk et al. 2001); (6) If present in the electron-lucent FCs, the "Christmas trees" should be identified by straightforward EM due to a high local concentration of RNA. In contrast, if present in the electron-dense DFCs, it is hard to distinguish such "Christmas trees" anyway since they are (part of) the DFCs.

Two recent papers (Gonzalez-Melendi et al. 2001; Koberna et al. 2002) add significant weight to the view that the "Christmas trees" are present in the DFCs and the DFC/FC border of both animal and plant somatic cells. I speculate that even though the FCs likely play an important architectural role in the arrangement of active ribosomal genes in animal somatic cells, they primarily serve as depots of silent genes. However, once a given gene is fully active in the form of the "Christmas tree", it automatically becomes, through the process of transcription, (part of) the DFC (Raška et al. 1989, 1995).

Three possible models of nucleoli in animal somatic cells have been recently listed by Huang (2002) with regard to the substructures involved in the pre-rRNA synthesis: the FCs, the FC/DFC border and finally the DFCs models. My opinion is that the fully active genes are not part of the FCs.

The FC/DFC border model (Huang 2002) in which the active genes line up along the border of the FCs and nascent transcripts are elongated into the DFCs requires a rather strong topological constraint. As far as I know, everybody agrees on compacted and contorted "Christmas trees" in situ, while this model requests a very special geometry for the nascent transcripts. The "Christmas trees" should then possess a laterally polarized form for which no hints are available (e.g., Scheer et al. 1997).

With regard to the documented model in which the DFCs are involved in pre-rRNA synthesis (Huang 2002), I prefer a modified model of animal somatic cell nucleoli. I am of the opinion that the fully active genes are (part of) the DFCs. However, the initiation sites of transcription of (some) active genes may well be found in the DFC/FC border zone, and in this sense, the DFC model is partially compatible with the FC/DFC border model. The DFC model for the arrangement of highly active ribosomal genes in animal somatic cell nucleoli does not explain why such genes are also spatially separated from the FCs in plant cells (Melčák et al. 1996; Gonzalez-Melendi et al. 2001). However, at least it identifies the DFCs as the nucleolar substructure containing the "Christmas trees" both in animal and plant somatic cells.

On a conciliatory note, if there are active ribosomal genes in somatic cells of higher eukaryotes that exhibit a much lower RNA polymerase I loading rate (encountered, for example, during the activation of a ribosomal gene or after the drug treatment) than that seen in the true "Christmas trees", i.e., there are just a few nascent transcripts associated with such genes, it may be that these genes are present in the FCs. However, in this report I consider exclusively the location of the "Christmas trees" under the physiological steady state situation encountered in metabolically active animal and plant somatic cells. Such cells have to posses highly expressed ribosomal genes as well.

I conclude that the pioneering autoradiographic study by Granboulan and Granboulan (1965) provides an accurate localization of the "Christmas trees" within the nucleoli. Most importantly, a simple remedy to this long-standing controversy is that the groups that basically use the same techniques and same cell types, but obtain adverse results, should make a collaborative project in order to settle this whole issue.

Acknowledgments. I.R. would like to thank Ann Beyer, Stanislav Fakan, Pierre-Emmanule Gleizes, Karel Koberna, Jan Malínský, Ulrich Scheer, Peter Shaw and David Staněk for discussion. Thanks are also due to Ann Beyer for the possibility to use and quote her unpublished data, Ulrich Scheer for the kind gift of the Christmas tree image, and all members of the Department of Cell Biology and Laboratory of Gene Expression for help and support. This work was supported by grants from the Grant Agency of the Czech Republic 304/00/1622, 304/01/0729, 304/02/0342 and 304/03/1121, from the Academy of Sciences of the Czech Republic IAA5039103 and AV0Z5039906, and from the Ministry of Education, Youth and Sports MSM: 111100003.

References

Anastassova-Kristeva M (1977) The nucleolar cycle in man. J Cell Sci 25:103–110

Andersen JS, Lyon CE, Fox AH, Leung AK, Lam YW, Steen H, Mann M, Lamond AI (2002) Directed proteomic analysis of the human nucleolus. Curr Biol 12:1–11

Bachellerie JP, Nicoloso M, Zalta JP (1977) Nucleolar chromatin in Chinese hamster ovary cells. Topographical distribution of ribosomal DNA sequences and isolation of ribosomal transcription complexes. Eur J Biochem 79:23–32

Bernardi R, Pandolfi PP (2003) The nucleolus: at the stem of immortality. Nature Medicine 9:24–25

Bernhard W (1969) A new staining procedure for electron microscopic cytology. J Ultrastruct Res 27:250–256

Beven AF, Lee R, Razaz M, Leader DJ, Brown JW, Shaw PJ (1996)The organization of ribosomal RNA processing correlates with the distribution of nucleolar snRNAs. J Cell Sci 109:1241–1251

Biberfeld P (1971) Morphogenesis in blood lymphocytes stimulated with phytohaemagglutinin (PHA). A light and electron microscopic study. Acta Pathol Microbiol Scand 223:1–70

Biggiogera M, Fakan S, Kaufmann SH, Black A, Shaper JH, Busch H (1989) Simultaneous immunoelectron microscopic visualization of protein B23 and C23 distribution in the HeLa cell nucleolus. J Histochem Cytochem 37:1371–1374

Biggiogera M, Burki K, Kaufmann SH, Shaper JH, Gas N, Amalric F, Fakan S (1990) Nucleolar distribution of proteins B23 and nucleolin in mouse preimplantation embryos as visualized by immunoelectron microscopy. Development 110:1263–1270

Biggiogera M, Malatesta M, Abolhassani-Dadras S, Amalric F, Rothblum LI, Fakan S (2001) Revealing the unseen: the organizer region of the nucleolus. J Cell Sci 114:3199–3205

Borovjagin AV, Gerbi SA (2001) Xenopus U3 snoRNA GAC-Box A' and Box A play distinct functional roles in rRNA processing. Moll Cell Biol 21:6210–6221

Brachet J (1940) La détection histochemique des acides pentosenucléiques. Compt Rend Soc Biol 133:88–90

Bunney TD, Watkins PA, Beven AF, Shaw PJ, Hernandez LE, Lomonossoff GP, Shanks M, Peart J, Drobak BK (2000) Association of phosphatidylinoditol 3-kinase with nuclear transcript sites in higher plants. Plant Cell 12:1679–1688

Busch H, Smetana K (1970) The Nucleolus. New York, London: Academic Press

Carmo-Fonseca M, Cuntha C, Custodio N, Carvalho C, Jordan P, Ferreira J, Parreira L (1996) The topography of chromosomes and genes in the nucleus. Exp Cell Res 229:247–252

Caspersson TO, Schultz J (1940) Ribonucleic acids in both nucleolus and cytoplasm and the function of the nucleolus. Proc Nat Acad Sci USA 26:507–515

Cavanaugh AH, Thompson EA Jr (1985) Hormonal regulation of transcription of rDNA: glucocorticoid effects upon initiation and elongation in vitro. Nucleic Acids Res 13:3357–3369

Chen D, Huang S (2001) Nucleolar components involved in ribosome biogenesis cycle between the nucleolus and nucleoplasm in interphase cells. J Cell Biol 153:169–176

Cheutin T, O'Donohue MF, Beorchia A, Vandelaer M, Kaplan H, Defever B, Ploton D, Thiry M (2002) Three-dimensional organization of active rRNA genes within the nucleolus. J Cell Sci 115:3297–3307

Chooi WY, Leiby KR (1981) An electrom microscopic method for localization of ribosomal proteins during transcription of ribosomal DNA: A method for studying protein assembly. Proc Natl Acad Sci USA 78:4823–4827

Chooi WY, Leiby KR (1983) Electron microscopic evidence for RNA polymerase loading at repeated sequences in non-transcribed spacers of D. virilis. Exp Cell Res 154:181–190

Christensen ME, Beyer AL, Walker B, LeStourgeon WM (1977) Identification of NG,NG dimethyl arginine in a nuclear protein from the lower eucaryote Physarum polycephalum homologous to the major proteins of mammalian 40S ribonucleoprotein particles. Biochem Biophys Res Comm 74:621–629

Clark MW, Yip ML, Campbell J, Abelson J (1990) SSB-1 of the yeast Saccharomyces cerevisiae is a nucleolar-specific, silver-binding protein that is associated with the snR10 and snR11 small nuclear RNAs. J Cell Biol 111:1741–1751

Cmarko D, Verschure PJ, Martin TE, Dahmus ME, Krause S, Fu XD, van Driel R, Fakan S (1999) Ultrastructural analysis of transcription and splicing in the cell nucleus after bromo-UTP microinjection. Mol Biol Cell 10:211–223

Cmarko D, Verschure PJ, Rothblum LI, Hernandez-Verdun D, Amalric F, van Driel R, Fakan S (2000) Ultrastructural analysis of nucleolar transcription in cells microinjected with 5-bromo-UTP. Histochem Cell Biol 113:181–187

Cogliati R, Gautier A (1973) Demonstration of DNA and polysaccharides using a new "Schiff type" reagent. CR Acad D 276:3041–3044

Conconi A, Widmer RM, Koller T, Sogo JM (1989) Two different chromatin structures coexist in ribosomal RNA genes throughout the cell cycle. Cell 57:753–761

Cook PR (1999) The organization of replication and transcription. Science 284:1790–1795

De Carcer G, Medina FJ (1999) Simultaneous localization of transcription and early processing markers allows dissection of functional domains in the plant cell nucleolus. J Struct Biol 128:139–151

Derenzini M, Thiry M, Goessens G (1990) Ultrastructural cytochemistry of the mammalian cell nucleolus. J Histochem Cytochem 38:1237–1256

Dragon F, Gallagher JE, Compagnone-Post PA, Mitchell BM, Porwancher KA, Wehner KA, Wormsley S, Settlage RE, Shabanowitz J, Osheim Y, Beyer AL, Hunt DF, Baserga SJ (2002) A large nucleolar U3 ribonucleoprotein required for 18S ribosomal RNA biogenesis. Nature 417:967–970

Dundr M, Raška I (1993) Nonisotopic ultrastructural mapping of transcription sites within the nucleolus. Exp Cell Res 208:275–281

Dundr M, Olson MO (1998) Partially processed pre-rRNA is preserved in association with processing components in nucleolus-derived foci during mitosis. Mol Biol Cell 9:2407–2422

Dundr M, Hoffmann-Rohrer U, Hu Q, Grummt I, Rothblum LI, Phair RD, Misteli T (2002) A kinetic framework for a mammalian RNA polymerase in vivo. Science 298:1623–1626

Fakan S (1986) Structural support for RNA synthesis in the cell nucleus. Methods Achiev Exp Pathol 12:105–140

Fath S, Milkereit P, Podtelejnikov AV, Bischler N, Schultz P, Bier M, Mann M, Tschochner H (2000) Association of yeast RNA polymerase I with a nucleolar substructure active in rRNA synthesis and processing. J Cell Biol 149:575–590

Fatica A, Tollervey D (2002) Making ribosomes. Curr Opin Cell Biol 14:313–318

Filipowicz W, Pogačic V (2002) Biogenesis of small nucleolar ribonucleoproteins. Curr Opin Cell Biol 14:319–327

Fischer D, Weisenberger D, Scheer U (1991) Assigning functions to nucleolar structures. Chromosoma 101:133–140

Franke WW, Scheer U, Spring H, Trendelenburg MF, Zentgraf H (1979) Organization of nucleolar chromatin. In: Busch H (eds) The cell nucleus, vol 7. Academic Press, New York, pp 49–95

French SL, Osheim YN, Cioci F, Nomura M, Beyer AL (2002) In exponentially growing Saccharomyces cerevisiae cells, ribosomal RNA synthesis is determined by the summed Pol I loading rate rather then by the number of active genes. Mol Cell Biol 23:1558–1568

Garcia-Blanco MA, Miller DD, Sheetz MP (1995) Nuclear spreads: I. Visualization of bipartite ribosomal RNA domains. J Cell Biol 128:15–27

Gas N, Escande ML, Stevens BJ (1985) Immunolocalization of the 100 kDa nucleolar protein during the mitotic cycle in CHO cells. Biol Cell 53:209–218

Geuskens M, Bernhard W (1966) Ultrastructural cytochemistry of the nucleolus. 3. The effect of actinomycin D on the metabolism of nucleolar RNA. Exp Cell Res 44:579–598

Goessens G (1976) High resolution autoradiographic studies of Ehrlich tumour cell nuclei. Exp Cell Res 100:88–94

Goessens G (1984) Nucleolar structure. Int Rev Cytol 87:107–158

Goessens G, Lepoint A (1979)The NORs: recent data and hypotheses. Biol Cell 35:211–220

Gonzalez-Melendi P, Testillano PS, Mena CG, Müller S, Raška I, Risueno MC (1998) Histones and DNA ultrastructural distribution in plant cell nucleus: a combination of immunogold and cytochemical methods. Exp Cell Res 242:45–59

Gonzalez-Melendi P, Beven AF, Boudonck K, Abranches R, Wells B, Dolan L, Shaw PJ (2000) The nucleus: a highly organized but dynamic structure. J Microsc 198:199–207

Gonzalez-Melendi P, Wells B, Beven AF, Shaw PJ (2001) Single ribosomal transcription units are linear, compacted Christmas trees in plant nucleoli. Plant J 27:223–233

Goodpasture C, Bloom SE (1975) Visualization of nucleolar organizer regions im mammalian chromosomes using silver staining. Chromosoma 53:37–50

Grainger RM, Maizels N (1980) Dictyostelium ribosomal RNA is processed during transcription. Cell 20:619–623

Granboulan N, Granboulan P (1965) Cytochemie ultrastructural du nucleole:II etude des sites de synthese du RNA dans le nucleole et le noyau. Exp Cell Res 38:604–619

Gratzner HG (1982) Monoclonal antibody to 5-bromo- and 5-iododeoxyuridine: a new reagent for detection of DNA replication. Science 218:474–475

Griffiths G (1993) Fine structure immunocytochemistry. Springer, Berlin Heidelberg New York

Grummt I (1978) In vitro synthesis of pre-rRNA in isolated nucleoli. In: Busch H (eds) The cell nucleus, vol 5. Academic Press, New York, pp 373–412

Haaf T, Ward DC (1996) Inhibition of RNA polymerase II transcription causes chromatin decondensation, loss of nucleolar structure, and dispersion of chromosomal domains. Exp Cell Res 224:163–173

Hadjiolov AA (1985) The nucleolus and ribosome biogenesis. In: Beerman AM, Goldstein L, Portrer KR, Sitte P (eds) Cell biology monographs. Springer, Berlin Heidelberg New York, pp 1–263

Hernandez-Verdun D (1991) The nucleolus today. J Cell Sci 99:465–471

Hernandez-Verdun D, Hubert J, Bourgeois C, Bouteille M (1978) Ultrastructural identification of the nucleolus organizer by the silver staining technic. CR Acad Sci D 287:1421–1423

Howell WM (1982) Selective staining of nucleolus organizer regions (NORs). In: Busch H (ed) The cell nucleus, vol 11. Academic Press, New York, pp 90–142

Hozák P, Cook PR, Schöfer C, Mosgöller W, Wachtler F (1994) Site of transcription of ribosomal RNA and intranucleolar structure in HeLa cells. J Cell Sci 107:639–648

Huang S (2002) Building an efficient factory: where is pre-rRNA synthesized in the nucleolus? J Cell Biol 157:739–741

Hügle B, Hazan R, Scheer U, Franke WW (1985) Localization of ribosomal protein S1 in the granular component of the interphase nucleolus and its distribution during mitosis J Cell Biol 100:873–886

Humbel BM, de Jong MD, Müller WH, Verkleij AJ (1998) Pre-embedding immunolabeling for electron microscopy: an evaluation of permeabilization methods and markers. Microsc Res Tech 42:43–58

Iben S, Tschochncr H, Bier M, Hoogstraten D, Hozák P, Egly JM, Grummt I (2002) TFIIH plays an essential role in RNA polymerase I transcription. Cell 109:297–306

Iborra FJ, Jackson DA, Cook PR (1998) The path of transcripts from extra-nucleolar synthetic sites to nuclear pores: transcripts in transit are concentrated in discrete structures containing SR proteins. J Cell Sci 111:2269–2282

Jackson DA, Hassan AB, Errington RJ, Cook PR (1993) Visualization of focal sites of transcription within human nuclei. EMBO J 12:1059–1065

Jackson DA, Iborra FJ, Manders EM, Cook PR (1998) Numbers and organization of RNA polymerases, nascent transcripts and transcription units in HeLa nuclei. Mol Biol Cell 9:1523–1236

Jacobson MR, Cao LG, Wang YL, Pederson T (1995) Dynamic localization of RNase MRP RNA in the nucleolus observed by fluorescentRNA cytochemistry in living cells. J Cell Biol 131:1649–1658

Jimenez-Garcia LF, Segura-Valdez ML, Ochs RL, Echeverria OM, Vazquez-Nin GH, Busch H (1993) Electron microscopic localization of ribosomal DNA in rat liver nucleoli by nonisotopic in situ hybridization. Exp Cell Res 207:220–225

Jordan EG, McGovern JH (1981) The quantitative relationship of the fibrillar centres and other nucleolar components to changes in growth conditions, serum deprivation and low doses of actinomycin D in cultured diploid human fibroblats (strain MRC-5). J Cell Sci 4:3–15

Jordan P, Mannervik M, Tora L, Carmo-Fonseca M (1996) In vivo evidence that TATA-binding protein SL1 colocalizes with UBF and RNA polymerase I when rRNA synthesis is either active or inactive. J Cell Biol 133:225–234

Junera HR, Masson C, Geraud G, Hernandez-Verdun D (1995) The three-dimensional organization of ribosomal genes and the architecture of the nucleoli vary with G1, S and G2 phases. J Cell Sci 108:3427–3441

Kass S, Sollner-Webb B (1990) The first re-rRNA- processing event occurs in a large complex: analysis by gel retardation. Mol Cell Biol 10:4920–4931

Kass S, Tyc K, Steitz JA, Sollner-Webb B (1990) The U3 small nucleolar ribonucleoprotein functions in the first step of preribosomal RNA processing. Cell 60:897–908

Kill IR (1996) Localization of the Ki-67 antigen within the nucleolus. Evidence for a fibrillarin-deficient region of the dense fibrillar component. J Cell Sci 109:1253–1263

Koberna K, Staněk D, Malínský J, Eltsov M, Pliss A, Čtrnáctá V, Cermanová Š, Raška I (1999) Nuclear organization studied with the help of a hypotonic shift: its use permits hydrophilic molecules to enter into living cells. Chromosoma 108:325–335

Koberna K, Staněk D, Malínský J, Čtrnáctá V, Cermanová Š, Novotná J, Kopský V, Raška I (2000) In situ fluorescence visualization of bromouridine incorporated into newly transcribed nucleolar RNA. Acta Histochem 102:15–20

Koberna K, Malínský J, Pliss A, Mašata M, Večeřová J, Fialová M, Bednár J, Raška I (2002) Ribosomal genes in focus: new transcripts label the dense fibrillar components and form clusters indicative of "Christmas trees" in situ. J Cell Biol157:743–748

Lazdins IB, Delannoy M, Sollner-Webb B (1997) Analysis of nucleolar transcription and processing domains and pre-rRNA movements by in situ hybridization. Chromosoma 105:481–495

Léger-Silvestre I, Trumtel S, Noaillac-Depeyre J, Gas A (1999) Functional compartmentalization of the nucleus in the budding yeast Saccharomyces cerevisiae. Chromosoma 108:103–113

Lewis H, Birnstiel M, Brown D, Gall J, Penman S, Perry R, Vincent W (1966) Panel on ribosome biogenesis. Natl Cancer Inst Monogr 23:547–561

Liau MC, Perry RP (1969) Ribosome precursor particles in nucleoli. J Cell Biol 42:272–283

Long EO, Dawid IB (1980) Repeated genes in eukaryotes. Annu Rev Biochem 49:727–764

Mais C, Scheer U (2001) Molecular architecture of the amplified nucleoli of Xenopus oocytes. J Cell Sci 114:709–718

Malínský J, Koberna K, Bednár J, Štulík J, Raška I (2002) Searching for active ribosomal genes in situ: light microscopy in light of the electron beam. J Struct Biol 140:227–231

Marinozzi V, Bernhard W (1963) Présence dans le nucléole de deux types de ribonucleoprotéines morphologiquement distinctes. Exp Cell Res 32:595–598

Masson C, Bouniol C, Famproix N, Szollosi MS, Debey P, Hernandez-Verdun D (1996) Conditions favoring RNA polymerase I transcription in permeabilized cells. Exp Cell Res 226:114–125

Matera G, Tycowski K, Steitz J, Ward D (1994) Organization of snoRNPs by fluorescence in situ hybridization and immunocytochemistry. Mol Biol Cell 5:1289–1299

Meier UT, Blobel G (1992) Nopp140 shuttles on tracks between nucleolus and cytoplasm. Cell 70:127–138

Meier UT, Blobel G (1994) NAP57, a mammalian nuclelar protein with a putative homolog in yeast and bacteria. J Cell Biol 127:1505–1514

Melčák I, Risueno MC, Raška I (1996) Ultrastructural nonisotopic mapping of nucleolar transcription sites in onion protoplasts. J Struct Biol 116:253–263

Mena CG, Testillano PS, Gonzales-Melendi P, Gorab E, Risueno MC (1994) Immunoelectron microscopy of RNA combined with nucleic acid cytochemistry in plant nucleoli. Exp Cell Res 212:393–408

Miller OL Jr, Beatty BR (1969) Visualization of nucleolar genes. Science 164:955–957

Moreno FJ, Hernandez-Verdun D, Masson C, Bouteille M (1985) Silver staining of the nucleolar organizer regions (NORs) on Lowicryl and cryo-ultrathin sections. J Histochem Cytochem 33:389–399

Mosgöller W, Schöfer C, Derenzini M, Steiner M, Maier U, Wachtler F (1993) Distribution of DNA in human Sertoli cell nucleoli. J Histochem Cytochem 41:1487–1493

Mosgöller W, Schöfer C, Wesierska-Gadek J, Steiner M, Müller M, Wachtler F(1998) Ribosomal gene transcription is organized in foci within nucleolar components. Histochem Cell Biol 109:111–118

Mosgöller W, Schöfer C, Steiner M, Sylvester JE, Hozák P (2001) Arrangement of ribosomal genes in nucleolar domains revealed by detection of "Christmas tree" components. Histochem Cell Biol 116:495–505

Mougey EB, O'Reilly M, Osheim Y, Miller OL Jr, Beyer A, Sollner-Webb B (1993) The terminal balls characteristic of eukaryotic rRNA transcription units in chromatin spreads are rRNA processing complexes. Genes Dev 7:1609–1619

Noaillac-Depeyre J, Dupont MA, Tichadou JL, Gas N (1989) The effect of adenosine analogue (DRB) on a major nucleolar phosphoprotein nucleolin. Biol Cell 67:27–35

Ochs RL, Smetana K (1989) Fibrillar center distribution in nucleoli of PHA-stimulated human lymphocytes. Exp Cell Res 184:552–557

Ochs RL, Lischwe MA, Spohn WH, Busch H (1985) Fibrillarin: a new protein of the nucleolus identified by autoimmune sera. Biol Cell 54:123–133

Olmedilla A, Testillano PS, Vincente O, Delseny M, Risueno MC (1993) Ultrastructural rRNA localization in plant cell nucleoli. RNA/RNA in situ hybridization, autoradiography and cytochemistry. J Cell Sci 106:1333–1346

Olson MO, Dundr M, Szebeni A (2000) The nucleolus: an old factory with unexpected capabilities. Trends Cell Biol 10:189–196

Olson MOJ, Hingorani K, Szebeni A (2002) Conventional and noncoventional roles of the nucleolus. Int Rev Cytol 219:199–266

Osheim YN, Mougey EB, Windle J, Anderson M, O'Reilly M, Miller OL, Beyer A, Sollner-Webb B (1996) Metazoan rDNA enhancer acts by making more genes transcriptionally active. J Cell Biol 133:943–954

Prior CP, Cantor CR, Johnson EM, Littau VC, Allfrey VG (1983) Reversible changes in nucleosome structure and histone H3 accessibility in transcriptionally active and inactive states of rDNA chromatin. Cell 34:1033–1042

Puvion-Dutilleul F, Bachellerie JP, Bernadac, Zalta JP (1977) Transcription complexes in subnuclear fractions isolated from mammalian cells: ultrastructural study. CR Acad Sci D 284:663–666

Puvion-Dutilleul F, Bachellerie JP, Puvion E (1991a) Nucleolar organization of HeLa cells as studied by in situ hybridization. Chromosoma 100:395–409

Puvion-Dutilleul F, Mazan S, Nicoloso M, Christensen ME, Bachellerie JP (1991b) Localization of U3 RNA molecules in nucleoli of HeLa and mouse 3T3 cells by high resolution in situ hybridization. Eur J Cell Biol 56:178–186

Puvion-Dutilleul F, Mazan S, Nicoloso M, Pichard E, Bachellerie JP, Puvion E (1992) Alterations of nucleolar ultrastructure and ribosome biogenesis by actinomycin D. Implications for U3 snRNP function. Eur J Cell Biol 58:149–162

Puvion-Dutilleul F, Puvion E, Bachellerie JP (1997) Early stages of pre-rRNA formation within the nucleolar ultrastructure of mouse cells studied by in situ hybridization with a 5'ETS leader probe. Chromosoma 105:496–505

Raška I, Dundr M (1993) Compartmentalization of the cell nucleus: case of the nucleolus. In: Stahl A, Luciano J, Vagner-Capodano A (eds) Chromosomes today, vol 11. Alen and Unwin, London, pp 101–119

Raška I, Armbruster BL, Frey JR, Smetana K (1983a) Analysis of ring-shaped nucleoli in serially sectioned human lymphocytes. Cell Tissue Res 234:707–711

Raška I, Rychter Z, Smetana K (1983b) Fibrillar centres and condensed nucleolar chromatin in resting and stimulated human lymphocytes. Z Mikrosk Anat Forsch 97:15–32

Raška I, Reimer G, Jarnik M, Kostrouch Z, Raška K Jr (1989) Does the synthesis of ribosomal RNA take place within nucleolar fibrillar centers or dense fibrillar components? Biol Cell 65:79–82

Raška I, Ochs RL, Salamin-Michael L (1990) Immunocytochemistry of the cell nucleus. Electron Microsc Rev 3:301–353

Raška I, Dundr M, Koberna K (1992) Structure-function subcompartments of the mammalian cell nucleus as revealed by the electron microscopic affinity cytochemistry. Cell Biol Int Rep 16:771–789

Raška I, Dundr M, Koberna K, Melčák I, Risueno MC, Török I (1995) Does the synthesis of ribosomal RNA take place within nucleolar fibrillar centers or dense fibrillar components? A critical appraisal. J Struct Biol 114:1–22

Recher L, Whitescarver J, Briggs L (1969) The fine structure of a nucleolar constituent. J Ultrastruct Res 29:1–14

Reeder RH (1999) Regulation of RNA polymerase I transcription in yeast and vertebrates. Prog Nucleic Acid Res Mol Biol 62:293–327

Reimer G, Raška I, Tan EM, Scheer U (1987) Human autoantibodies: probes for nucleolus structure and function. Virchows Arch B 54:131–143

Reimer G, Raška I, Scheer U, Tan EM (1988) Immunolocalization of 7-2-ribonucleoprotein in the granular component of the nucleolus. Exp Cell Res 176:117–128

Rendón MC, Rodrigo RM, Goenechea LG, Garcia-Herdugo G, Valdivia MM, Moreno FJ (1992). Characterization and immunolocalization of a nucleolar antigen with anti-NOR serum in HeLa cells. Exp Cell Res 200:393–403

Risueno MC, Medina FJ (1986) The nucleolar structure in plant cells. Revis Biol Celular 7:1–154

Risueno MC, Testillano PS (1994) Cytochemistry and immunocytochemistry of nucleolar chromatin in plants. Micron 25:331–360

Roussel P, Andre C, Masson C, Geraud G, Hernandez-Verdun D (1993) Localization of the RNA polymerase I transcription factor hUBF during the cell cycle. J Cell Sci 104:327–337

Royal A, Simard R (1975) RNA synthesis in the ultrastructural and biochemical components of the nucleolus of Chinese hamster ovary cells. J Cell Biol 66:577–585

Sandmeier JJ, French S, Osheim Y, Cheung WL, Gallo CM, Beyer AL, Smith JS (2002) RPD3 is required for the inactivation of yeast ribosomal DNA genes in stationary phase. EMBO J 21:4959–4968

Santoro R, Grummt I (2001) Molecular mechanisms mediating methylation-dependent silencing of ribosomal gene transcription. Mol Cell 8:719–725

Santoro R, Li J, Grummt I (2002) The nucleolar remodeling complex NoRC mediates heterochromatin formation and silencing of ribosomal gene transcription. Nat Genet 32:393–396

Savino R, Gerbi SA (1990) In vivo disruption of Xenopus U3 snRNA affects ribosomal RNA processing. EMBO J 9:2299–2308

Schäfer T, Strauss D, Petfalski E, Tollervey D, Hurt E (2003) The path from nucleolar 90S to cytoplasmic 40S pre-ribosomes. EMBO J 22:1370–1380

Scheer U (1978) Changes of nucleosome frequency in nuclear and non-nuclear chromatin as a function of transcription: an electron microscopic study. Cell 13:535–549

Scheer U (1987) Contributions of electron microscopic spreading preparations ("Miller spreads") to the analysis of chromosome structure. Res Prob Cell Differ 14:147–171

Scheer U, Benavente R (1990) Functional and dynamic aspects of the mammalian nucleolus. Bioessays 12:14–21

Scheer U, Hock R (1999) Structure and function of the nucleolus. Curr Opin Cell Biol 11:385–390

Scheer U, Raška I (1987) Immunocytochemical localization of RNA polymerase I in the fibrillar centers of nucleoli. In: Stahl A, Luciano J, Vagner-Capodano A (eds) Chromosomes today, vol 9. Alen and Unwin, London, pp 284–294

Scheer U, Rose KM (1984) Localization of RNA polymerase I in interphase cells and mitotic chromosomes by light and electron microscopic immunocytochemistry. Proc Natl Acad Sci USA 81:1431–1435

Scheer U, Zentgraf H, Sauer HW (1981) Different chromatin structures in Physarum polycephalum: a special form of transcriptionally active chromatin devoid of nucleosomal particles. Chromosoma 84:279–290

Scheer U, Xia B, Merkert H, Weisenberger D (1997) Looking at Christmas trees in the nucleolus. Chromosoma 105:470–480

Scherl A, Couté Y, Déon C, Callé A, Kindbeiter K, Sanchez JC, Greco A, Hochstrasser D, Diaz JJ (2002) Functional proteomic analysis of human nucleolus. Mol Biol Cell 13:4100–4109

Scherrer K, Latham H, Darnell JE (1963) Demonstration of an unstable RNA and of a precursor to ribosomal RNA in HeLa cell. Proc Natl Acad Sci USA 49:240–248

Schmidt-Zachmann MS, Hügle-Dorr B, Franke WW (1987) A constitutive nucleolar protein identified as a member of the nucleoplasmin family. EMBO J 6:1881–1890

Schöfer C, Müller M, Leitner MD, Wachtler F (1993) The uptake of uridine in the nucleolus occurs in the dense fibrillar component. Immunogold localization of incorporated digoxigenin-UTP at the electron microscopic level. Cytogenet Cell Genet 64:27–30

Schwarzacher HG, Mikelsaar AV, Schnedl W (1978) The nature of the Ag-staining of nucleolus organizer regions. Electron- and light-microscopic studies on human cells in interphase, mitosis, and meiosis. Cytogenet Cell Genet 20:24–39

Shaw PJ, Jordan EG (1995) The nucleolus. Annu Rev Cell Dev Biol 11:93–121

Shaw PJ, Highett MI, Beven AF, Jordan EG (1995) The nucleolar architecture of polymerase I transcription and processing. EMBO J 14:2896–2906

Sinclair GD, Brasch K (1978) The reversible action of alpha-amanitin on nuclear structure and molecular composition. Exp Cell Res 111:1–14

Sirri V, Roussel P, Hernandez-Verdun D (2000) The AgNOR proteins: qualitative and quantitative changes during the cell cycle. Micron 31:121–126

Smetana K, Busch H (1974) The nucleolus and nucleolar DNA. In: Busch H (eds) The cell nucleolus, vol 1. Academic Press, New York, pp 73–147

Smetana K, Lejnar J, Potmesil M (1967) A note to the demonstration of DNA in nuclei of blood cells in smea preparations. Folia Haematol 88:305–317

Smetana K, Freireich EJ, Busch H (1968) Chromatin structures in ring-shaped nucleoli of human lymphocytes. Exp Cell Res 52:112–128

Spector D, Ochs R, Busch H (1884) Silver staining, immunofluorescence, and immunoelectron microscopic localization of nucleolar phosphoproteins B23 and C23. Chromosoma 90:139–148

Staněk D, Kiss T, Raška I (2000) Pre-ribosomal RNA is processed in permeabilised cells at the site of transcription. Eur J Cell Biol 79:202–207

Staněk D, Koberna K, Pliss A, Malínský J, Mašata M, Večeřová J, Risueno MC, Raška I (2001) Non-isotopic mapping of ribosomal RNA synthesis and processing in the nucleolus. Chromosoma 110:460–470

Stierhof YD, Schwarz H (1989) Labeling properties of sucrose-infiltrated cryosections. Scanning Microsc Supl 3:35–46

Stierhof YD, Schwarz H, Frank H (1986) Transverse sectioning of plastic-embedded immunolabeled cryosections: morphology and permeability to protein A-colloidal gold complexes. J Ultrastruct Mol Struct Res 97:187–196

Stockert JC, Fernandez-Gomez ME, Sogo JM, Lopez-Saez JF (1970) Nucleolar segregation by adenosine 3'-deoxyriboside (cordycepin) in root-tip cells of Allium cepa. Exp Cell Res 59:85–89

Sullivan GJ, Bridger JM, Cuthbert AP, Newbold RF, Bickmore WA, McStay B (2001) Human acrocentric chromosomes with transcriptionally silent nucleolar organizer regions associate with nucleoli. EMBO J 20:2867–2874

Testillano PS, Sanchez-Pina MA, Olmedilla A, Ollacarizqueta MA, Tandler CJ, Risueno MC (1991) A specific ultrastructural method to reveal DNA: the NAMA-Ur. J Histochem Cytochem 39:1427–1438

Testillano PS, Gorab E, Risueno MC (1994) A new approach to map transcription sites at the ultrastructural level. J Histochem Cytochem 42:1–10

Thiry M (1991) In situ nick translation at the electron microscopy level: A tool for studying the location of Dnase I-sensitive regions within the cell. J Histochem Cytochem 39:871–874

Thiry M, Goessens G (1986) Ultrastructural study of the relationships between the various nucleolar components in Ehrlich tumour and HEp-2 cell nucleoli after acetylation. Exp Cell Res 164:232–242

Thiry M, Goessens G (1991) Distinguishing the sites of pre-rRNA synthesis and accumulation in Ehrlich tumor cell nucleoli. J Cell Sci 99:759–767

Thiry M, Müller S (1989) Ultrastructural distribution of histones within Ehrlich tumor cell nucleoli: a cytochemical and immunocytochemical study. J Histochem Cytochem 37:853–862

Thiry M, Thiry-Blaise L (1989) In situ hybridization at the electron microscope level: an improved method for precise localization of ribosomal DNA and RNA. Eur J Cell Biol 50:235–243

Thiry M, Lepoint A, Goessens G (1985) Re-evaluation of the site of transcription in Ehrlich tumour cell nucleoli. Biol Cell 54:57–64

Thiry M, Scheer U, Goessens G (1991) Localization of nucleolar chromatin by immunocytochemistry and in situ hybridization at the electron microscopy level. Electron Microsc Rev 4:85–110

Thiry M, Ploton D, Menager M, Goessens G (1993) Ultrastructural distribution of DNA within the nucleolus of various animal cell lines or tissues revealed by terminal deoxynucleotidyl transferase. Cell Tissue Res 271:33–45

Thiry M, Cheutin T, O'Donohue MF, Kaplan H, Ploton D (2000) Dynamics and three-dimensional localization of ribosomal RNA within the nucleolus. RNA 6:1750–1761

Thompson WF, Beven AF, Wells B, Shaw PJ (1997) Sites of rDNA transcription are widely dispersed through the nucleolus in Pisum sativum and can comprise single genes. Plant J 12:571–581

Trendelenburg MF (1974) Morphology of ribosomal RNA cistrons in oocytes of the water beetle, Dytiscus marginalis L. Chromosoma 48:119–135

Trendelenburg MF, Spring H, Scheer U, Franke WW (1974) Morphology of nucleolar cistrons in a plant cell, Acetabularia Mediterranea Proc Natl Acad Sci USA 71:3626–3630

Tröster H, Spring H, Meissner B, Schultz P, Oudet P, Trendelenburg MF (1985) Structural organization of an active, chromosomal nucleolar organizer region (NOR) identified by light microscopy, and subsequent TEM and STEM electron microscopy. Chromosoma 91:151–163

Trumtel S, Leger-Silvestre I, Gleizes PE, Teulieres F, Gas N (2000) Assembly and functional organization of the nucleolus: ultrastructural analysis of Saccharomyces cerevisiae mutants. Mol Biol Cell 11:2175–2189

Tuteja N, Beven AF, Shaw PJ, Tuteja R (2001) A pea homologue of human DNA helicase I is localised within the dense fibrillar component of the nucleolus and stimulated by phosphorylation with CK2 and cdc2 protein kinases. Plant J 25:9–17

Vandelaer M, Thiry M, Goessens G (1993) Ultrastructural distribution of DNA within the ring-shaped nucleolus of human resting T lymphocytes. Exp Cell Res 205:430–432

Von Schack ML, Fakan S, Villiger W (1991) Some applications of cryosubstitution in ultrastructral studies of the cell nucleus. Biol Cell 72:113–119

Wachtler F, Stahl A (1993) The nucleolus: a structural and functional interpretation. Micron Microsc Acta 24:473–505

Wachtler F, Schöfer C, Mösgoller W, Weipoltshammer K, Schwarzacher HG, Guichaoua M, Hartung M, Stahl A, Berge-Lefranc JL, Gonzalez I, Sylvester J (1992) Human ribosomal RNA gene repeats are localized in the dense fibrillar component of nucleoli: light and electron microscopic in situ hybridization in human Sertoli cells. Exp Cell Res 198:135–143

Wansink DG, Schul W, van der Kraan I, van Steensel B, van Driel R, de Jong L (1993) Fluorescent labeling of nascent RNA reveals transcription by RNA polymerase II in domains scattered throughout the nucleus. J Cell Biol 122:283–293

Warner JR (1999) The economics of ribosome biosynthesis in yeast. Trends Biochem Sci 24:437–440

Wei T, Baiqu H, Chunxiang L, Zhonghe Z (2003) In situ visualization of rDDA arrangement and its relationship with subnucleolar structural regions in Allium sativum cell nucleolus. J Cell Sci 116:1117–1125

Zatsepina OV, Dudnic OA, Chentsov YS, Thiry M, Spring H, Trendelenburg MF (1997) Reassembly of functional nucleoli following in situ unraveling by low-ionic-strength treatment of cultured mammalian cells. Exp Cell Res 233:155–168

Toxic RNA in the Nucleus: Unstable Microsatellite Expression in Neuromuscular Disease

Keith R. Nykamp[1] and Maurice S. Swanson[1]

1
Overview

A variety of subnuclear structures, including nucleoli and Cajal bodies, are involved in the normal synthesis and processing of various types of gene transcripts as well as the formation and remodeling of their associated ribonucleoprotein (RNP)-RNA complexes. In contrast to this normal synthesis and assembly process, the expression of some mutant genes results in the formation of aberrant intranuclear inclusions that may interfere with nuclear function. These genes contain repetitive DNA sequence motifs, microsatellite tri- to penta-nucleotide repeats that contract and expand in length due to errors in DNA replication, recombination and repair. When microsatellite expansions occur within the protein coding region a toxic gene product may be synthesized. Recent studies suggest that some of these mutant proteins interfere with the function of factors important for transcriptional regulation. Surprisingly, a group of autosomal dominant neuromuscular diseases, the myotonic dystrophies, are associated with expansions in untranslated regions, and expression of these unstable microsatellites results in the accumulation of mutant allele transcripts together with specific proteins in nuclear foci. Are these RNA foci pathogenic or protective and do they provide evidence for the existence of yet another subnuclear structure in wild-type cells? Here, we review unstable microsatellites and their effects on nuclear structure and function with particular emphasis on the hypothesis that the transcription of microsatellite expansions results in the production of toxic RNA.

[1]Department of Molecular Genetics and Microbiology and Powell Gene Therapy Center, University of Florida College of Medicine, 32610-0266, Gainesville, Florida, USA. Tel. +1 352 392 3082; Fax. +1 352 392 3133, e-mail: mswanson@ufl.edu

Progress in Molecular and Subcellular Biology
P. Jeanteur (Ed.): RNA Trafficking and Nuclear Structure Dynamics
© Springer-Verlag Berlin Heidelberg 2003

2
Unstable Microsatellites in Neurological and Neuromuscular Disease

2.1
Microsatellites: Definition and Potential Functions

DNA sequence analysis has revealed that at least 50% of the human genome is composed of repetitive sequences (Lander et al. 2001). Within this repeat population are transposon-derived or interspersed repeats (LINEs, SINES, LTR elements, DNA elements), pseudogenes, segmental duplications, large tandemly repeated sequences (centromeres, telomeres, ribosomal gene clusters) and simple sequence repeats (SSRs). SSRs are often referred to as microsatellites, which are relatively short in length and are usually defined as 1–6 base pairs (bp), but have also been described as 1–13 bp, in length (Tóth et al. 2000; Lander et al. 2001). Microsatellites have played an important role as genetic markers because they are highly polymorphic in the population, possibly due to errors during DNA replication, mismatch repair and recombination (Bowater and Wells 2001).

Although the majority of microsatellites may be nonfunctional, when sequence repeats are located within genes or their controlling elements they have the potential to influence gene expression. For example, SSRs may be binding sites for DNA-binding proteins that enhance or repress gene expression. At the RNA level, repeats within transcribed regions may fold into stable RNA structures that possess either intrinsic RNA-based functions or act as binding sites for RNA-binding proteins. In this review, we will primarily discuss microsatellite expansions that are associated with neurological and neuromuscular diseases.

2.2
Coding Region Expansions and Potential Effects on Transcriptional Regulation

Although the focus of this review is on RNA-mediated pathogenesis, unstable microsatellites within coding regions also cause a number of inherited diseases. There are currently 19 characterized neurological and neuromuscular diseases associated with unstable microsatellite expansion (Table 1; for recent reviews, see Margolis and Ross 2001; Orr 2001). The most common repeat expansion associated with disease is $(CAG)_n$, which is translated into a poly-glutamine (polyQ) tract. Expression of polyQ alone may be toxic, but clinical features associated with these disorders indicate that protein context is critical for disease-specific pathogenesis. Early work highlighted an intriguing correlation between pathogenesis and the formation of intranuclear inclusions of

Table 1. Characterized coding and non-coding region expansion diseases

Disease	Gene	Normal/expanded repeat	Proposed disease mechanisms
Coding			
Dentatorubral pallidoluysian atrophy (DRPLA)	*DRPLA*	$CAG_{3-36}/_{49-88}$	Toxic polyglutamine in atrophin-1
Huntington chorea	*HD*	$CAG_{6-35}/_{36-121}$	Toxic polyglutamine in huntingtin
Oculopharyngeal muscular dystrophy (OPMD)	*PABPN1*	$GCG_6/_{8-13}$	Toxic polyalanine in PABPN1
Spinobulbar muscular atrophy (SBMA)	*AR*	$CAG_{9-36}/_{38-62}$	Toxic polyglutamine in androgen receptor
Spinocerebellar ataxia (SCA) type 1	*SCA1*	$CAG_{6-44}/_{40-82}$	Toxic polyglutamine in ataxin-1
SCA type 2	*SCA2*	$CAG_{14-32}/_{33-77}$	Toxic polyglutamine in ataxin-2
SCA type 3	*MJD*	$CAG_{12-40}/_{55-86}$	Toxic polyglutamine in ataxin-3
SCA type 6	*CACNA1A*	$CAG_{4-18}/_{21-30}$	Toxic polyglutamine in α1A (calcium channelopathy)
SCA type 7	*SCA7*	$CAG_{7-17}/_{38-200}$	Toxic polyglutamine in ataxin-7
SCA type 17	*TBP*	$CAG_{29-42}/_{47-63}$	Toxic polyglutamine in TATA-binding protein
Non-coding			
5′-UTR			
Fragile XA	*FMR1/FMRP*	$CGG_{6-60}/_{230->1000}$	FMRP loss-of-function
Fragile XE	*FMR2*	$GCC_{7-35}/_{>200}$	FMR-2 loss-of-function
SCA type 12	*PPP2R2B*	$CAG_{7-28}/_{66-78}$	Elevated PPP2R2B expression, toxic PPP2R2B RNA (CAG repeat is in promoter or 5′-UTR)
3′-UTR			
Myotonic dystrophy type 1 (DM1)	*DMPK*	$CTG_{5-37}/_{50->2000}$	DMPK haploinsufficiency, altered *DMPK* locus structure, DMPK isoform imbalance, toxic DMPK RNA
SCA type 8	*SCA8*	$CTG_{16-91}/_{107-127}$	Toxic SCA8 RNA, altered KLHL1 regulation
Intron			
Friedreich's ataxia	*FRDA*	$GAA_{8-22}/_{120-1700}$	Frataxin loss-of-function
Myotonic dystrophy Type 2 (DM2)	*ZNF9*	$CCTG_{5-9}/_{75-~11,000}$	Toxic ZNF9 RNA
SCA10	*SCA10*	$ATTCT_{10-22}/_{750-4500}$	Toxic SCA10 RNA
Unknown			
Huntington disease-like 2	*JPH3*	$CTG_{7-26}/_{44-57}$	Toxic JPH3 RNA

mutant proteins (reviewed in Orr 2001). However, more recent studies have supported the possibility that the soluble pool of polyQ protein is pathogenic, possibly because it competes for transcription factor binding with the chromatin-associated transcriptional machinery (Margolis and Ross 2001; Freiman and Tjian 2002). Therefore, the formation of intranuclear, and perhaps inaccessible, mutant protein inclusions could be viewed as a protective response of the cell to reduce the soluble pool. Since intranuclear polyQ inclusions are also ubiquitinated, other investigators have speculated that disease is due to inhibition of the ubiquitin-proteasome system by these very unusual mutant proteins (Bence et al. 2001; Waelter et al. 2001).

Another type of coding region expansion occurs in the neuromuscular disease oculopharyngeal muscular dystrophy (OPMD). A pathological hallmark of OPMD is the appearance of intranuclear filamentous inclusions within muscle fiber nuclei. However, these inclusions are present in <10% of OPMD patient myonuclei (Becher et al. 2000; Uyama et al. 2000). The molecular basis of OPMD is the expansion of a $(GCG)_n$ repeat in the *PABPN1* gene, which encodes the nuclear poly(A) tail binding protein (Brais et al. 1998). These repeats increase from 6 in normal individuals to 8–13 repeats in affected patients resulting in an expansion in a PABPN1 amino terminal alanine tract from 10 to 12–17 residues (nonGCG codons bordering the repeat encode the additional four alanines). Interestingly, the PABPN1 protein is an essential mRNA polyadenylation factor, and is therefore ubiquitously expressed in tissues, even though OPMD pathology is mainly restricted to skeletal muscle. In vitro studies with the mutant protein indicate that alanine tract expansion results in an insoluble form of PABPN1 which accumulates in intranuclear filamentous inclusions together with poly(A)$^+$ RNA (Becher et al. 2000; Calado et al. 2000). It is not clear why pathology is relatively specific to certain muscles since PABPN1-enriched inclusions are absent in ≥90% of OPMD myonuclei and there is no obvious relationship between the severity of muscle damage and frequency of PABPN1-positive aggregates (Becher et al. 2000; Uyama et al. 2000). Surprisingly, polyadenylation is not affected in OPMD myoblasts, but PABPN1 may also regulate muscle-specific transcription by interacting with Ski-interacting protein (SKIP) and MyoD (Calado et al. 2000; Kim et al. 2001).

In summary, both polyglutamine and polyalanine expansions appear to result in the synthesis of mutant proteins that have the potential to bind, and possibly sequester, key components of nuclear regulatory machineries. The assembly of these complexes may lead to further aggregation and the appearance of intranuclear inclusions, but the onset of disease does not require the formation of these structures. As we discuss in the following section, related models have been introduced for the RNA-mediated diseases with the formation of RNA-protein intranuclear inclusions, sequestration of factors required for normal alternative pre-mRNA processing and impairment of the nuclear RNA turnover machinery.

2.3
Myotonic Dystrophy and RNA-Mediated Pathogenesis

Not all microsatellites are located within coding regions. Unstable microsatellites are also found in introns as well as 5'-UTRs and 3'-UTRs. Repeats in the normal size range probably play important roles as regulatory RNA elements in alternative pre-mRNA splicing as well as mRNA trafficking, translation and turnover. Expansion of these noncoding microsatellites can lead to transcriptional repression and loss of the affected gene product, as is the case with Fragile X and Freidreich's ataxia (Table 1). In contrast, transcription of microsatellite expansions may result in a toxic RNA. Since evidence for RNA-mediated disease pathogenesis first arose during investigations on the molecular etiology of myotonic dystrophy type 1 (DM1), we will review what is known about this neuromuscular disease and then discuss the experimental data that support the toxic RNA theory.

DM1 is the most prevalent adult-onset muscular dystrophy, but this disease is unusual because it is caused by a $(CTG)_n$ expansion in the 3'-UTR of the *DMPK* gene (Harper 2001). One of the intriguing aspects of DM1 is the variable clinical phenotype and the number of tissues affected. Although characteristic features of this disease are primarily muscular, including myotonia or muscle hyperexcitability and myopathy or muscle wasting, other abnormalities include a distinctive rosette-type posterior cataract, premature frontal balding, testicular atrophy, insulin resistance and cognitive impairment (Harper 2001). Because DM1 is an autosomal dominant disorder caused by a noncoding trinucleotide repeat, early proposals suggested that disease was due to *DMPK* haploinsufficiency or to the effects of $(CTG)_n$ expansion on the linked genes *DMWD* and *SIX5* (reviewed in Mankodi and Thornton 2002; Ranum and Day 2002). Indeed, reduced *DMPK* expression in DM1 patient muscle and muscle-derived myoblasts has been reported although when protein levels are normalized to type 1 muscle fiber controls, DMPK is actually elevated (1.2–1.6-fold) in adult DM1 and only modestly decreased in CDM patients (Narang et al. 2000). Moreover, the haploinsufficiency model suggests that loss-of-function truncation or missense mutations in the *DMPK* coding region should exist, but this type of mutant *DMPK* allele has not been reported. Knockout *Dmpk* models have also been unsuccessful in replicating some characteristic skeletal muscle defects associated with DM1 disease. Neither $Dmpk^{+/-}$ nor $Dmpk^{-/-}$ mice are affected by myotonia although homozygous knockout mice do display DM1-related cardiac conduction abnormalities and late onset progressive skeletal myopathy (Jansen et al. 1996; Reddy et al. 1996; Berul et al. 1999, 2000). The *DMPK* gene is bordered by *DMWD* and *SIX5* and $(CTG)_n$ expansion causes a twofold decrease in transcript levels from both of these genes. Both $Six5^{+/-}$ and $Six5^{-/-}$ mice develop cataracts (Klesert et al. 2000; Sarkar et al. 2000). However, the type and distribution of this mutant cataract is quite different from the rosette-like cataract characteristic of DM1 (Ranum and Day 2002). The discovery of a novel *DMPK* exon 16 (E16) led to a third, or isoform

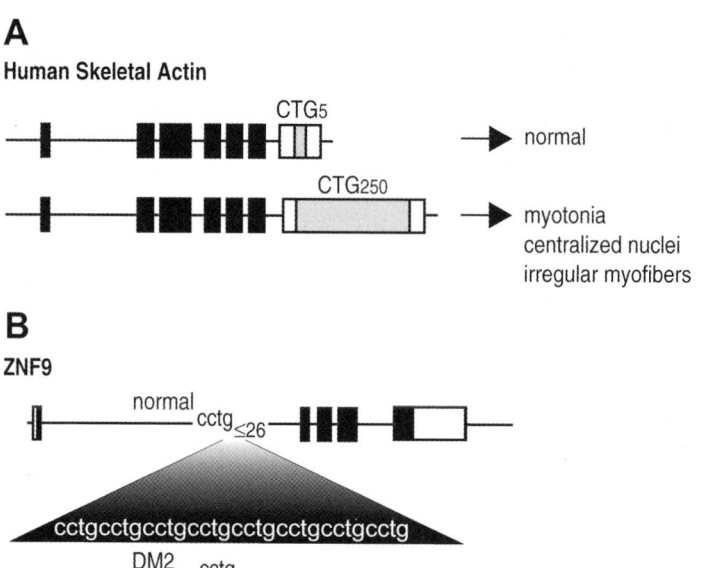

Fig. 1. Allele-independent pathogenesis of CTG-rich repeats. **A** Structure of the human skeletal actin (HSA) transgene used to construct a transgenic model for DM1. The structure of the HSA_{SR} (CTG$_5$) and HSA_{LR} (CTG$_{250}$) transgenes are illustrated. Exons (*black boxes* for coding regions, *open boxes* for untranslated regions), introns (*thin lines*) and the 3′-UTR insertion sites for the CTG$_5$ and CTG$_{250}$ repeats (*grey boxes*) are shown. Note that only transgenic mice expressing the HSA_{LR} transgene develop DM1-related muscle abnormalities including myotonia. **B** Structure of ZNF9 and the (CCTG)$_n$ expansion in the first intron which causes myotonic dystrophy type 2 (DM2). The ZNF9 gene is composed of five exons (*black/open boxes*) with a (CCTG)$_n$ repeat in the first intron which is ≤26 repeats in length in the normal population, but expands to 75– ~11,000 repeats in individuals with DM2 disease. The (CCTG)$_n$ repeat is located within a more complex region composed of several additional repeats ([TG]$_{14–25}$[TCTG]$_{4–10}$**[CCTG]**$_n$G/ TCTG[CCTG]$_{1–7}$ TCTG[CCTG]$_{4–8}$ in the normal population with the tetranucleotide expanded in DM2 highlighted in *bold* Liquori et al. 2001)

imbalance, model in which pre-mRNA splicing generates a DMPK mRNA in which the (CTG)$_n$ repeat is in the intron upstream of E16 (Tiscornia and Mahadevan 2000). Because this mRNA exits the nucleus efficiently while mutant DMPK mRNAs with the (CTG)$_n$ in the 3′-UTR are impaired for export (see below), (CTG)$_n$ expansion should increase the relative level of the DMPK E16 protein isoform. This prediction has not been tested directly.

All of the above disease models (*DMPK* haploinsufficiency, chromatin structure alteration in the *DMWD/DMPK/SIX5* region, DMPK isoform imbalance) depend on variations in the expression of the *DMPK* gene or closely linked genes. However, recent studies demonstrate that the (CTG)$_n$ repeat irrespective of gene context is pathogenic. For example, transgenic mice that express a human skeletal actin (HSA) gene with a long (CTG)$_{250}$ repeat positioned in the 3′-UTR (HSA_{LR}) develop myotonia while mice expressing a shorter HSA-(CTG)$_5$ repeat (HSA$_{SR}$) do not (Mankodi et al. 2000; Fig. 1A). The skeletal muscles of HSA$_{LR}$ mice are also structurally abnormal and show features characteristic of DM1 including centralized nuclei and fiber size heterogeneity. Another DM1 transgenic mouse model, created using a 45 kb fragment containing the *DMPK* gene from a DM1 patient and expressing ~300 CTG repeats, not only develops myotonia and myonuclear RNA foci, but also shows DM1-like isoform alterations in the neuronal protein tau (Seznec et al. 2001). Interestingly, knock-in mice that express shorter CTG repeat expansions, generated by replacing mouse *Dmpk* exons 13–15 with the corresponding human *DMPK* exons but carrying a (CTG)$_{84}$ repeat in exon 15, do not develop a DM1-like phenotype (van den Broek et al. 2002).

Additional strong support for the concept of RNA-mediated pathogenesis was also provided by the remarkable discovery that myotonic dystrophy type 2 (DM2) is caused by the expansion of a CTG-related repeat, (CCTG)$_n$, in the first exon of the *ZNF9* gene (Liquori et al. 2001; Figs. 1B and 2A). DM1 and DM2 share many clinical features, including myotonia, weakness in facial and neck muscles, posterior iridescent cataract, cardiac conduction defects and progressive cardiomyopathy (Ranum and Day 2002). Since expression of (CTG)$_n$ and (CCTG)$_n$ repeats cause similar diseases, what is the common pathogenic event? One possibility is that these repeats act at the DNA level and larger repeats trigger global alterations in chromatin structure and/or sequester (CTG)$_n$ or (CCTG)$_n$ DNA-binding proteins that are required for normal muscle development and maintenance. In the following section, we will discuss the alternative possibility that disease pathogenesis is mediated by RNA.

3
RNA-Mediated Disease

3.1
Nuclear RNA Foci in Myotonic Dystrophy

The hypothesis that *DMPK* mutant allele transcripts are toxic was initially proposed as a result of studies on the inhibitory effect of the expansion mutation on the levels of both normal and mutant DMPK mRNAs in the poly(A)$^+$ RNA fraction, but not from the total RNA pool (Wang et al. 1995). This result suggested that the DM1 (CUG)$_n$ expansion did not affect transcription of DMPK pre-mRNA from either the normal or mutant *DMPK* gene, but had a trans-dominant effect on the accumulation of normal DMPK poly(A)$^+$ RNA. Subsequent work failed to confirm the loss of normal polyadenylated DMPK mRNA in DM1 fibroblasts or myoblasts (Davis et al. 1997; Hamshere et al. 1997). Nevertheless, the toxic RNA theory gained timely support from fluorescence in situ hybridization (FISH) studies which demonstrated that transcripts from the mutant *DMPK* allele accumulated in the nucleus within foci (Taneja et al. 1995). This important study concluded that normal *DMPK* allele tran-

A

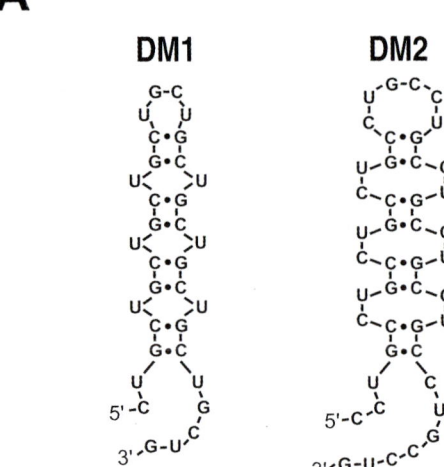

B

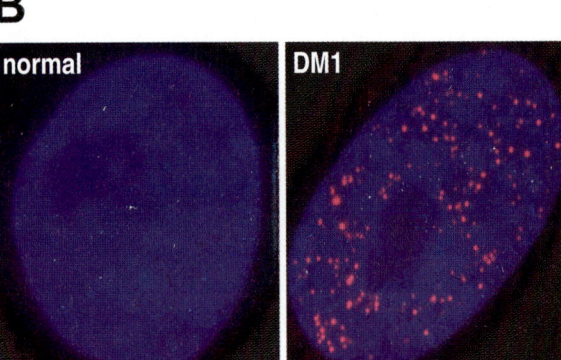

Fig. 2. Expansion of either CTG (DM1) or CCTG (DM2) repeats leads to the formation of dsRNA, appearance of mutant RNA nuclear foci and myotonic dystrophy. **A** Predicted structures (Zuker et al. 1999) of $(CUG)_{11}$ and $(CCUG)_{11}$ RNA repeats. Longer repeats are predicted to form extended RNA hairpins with dsCUG significantly more stable than dsCCUG. **B** Numerous RNA foci in DM1, but not in normal cells. Either normal or DM1 fibroblasts were converted to "myoblasts" by recombinant adenovirus-MyoD infection followed by FISH detection of *DMPK* mutant allele transcripts using a CY3-conjugated $(CAG)_{10}$ hybridization probe (*red dots*). Chromosomal DNA is stained with DAPI (*blue*). Normal DMPK transcripts in this fibroblast line express a *DMPK* gene with only five CTG repeats that does not hybridize to the $(CAG)_{10}$ probe under the stringency conditions employed

scripts do not accumulate in foci and thus the trans-dominant effect might result from a general inhibition of nuclear functions by mutant DMPK RNA foci. Subsequent studies have confirmed that the number and size of these nuclear foci is dependent on the expression level of the *DMPK* mutant allele with more numerous and larger foci in DM1 myoblasts than in fibroblasts. Induction of myoblast fusion by infection with a retrovirus or adenovirus expressing MyoD leads to the formation of >100 foci per myonucleus (Fig. 2B; Davis et al. 1997; Miller et al. 2000). Subcellular fractionation studies confirm that mutant *DMPK* gene transcripts are retained in the nucleus in association with the nuclear matrix (Davis et al. 1997; Hamshere et al. 1997). Of particular interest is the observation that nuclear DMPK RNA retention shows a threshold between 80 and 400 CUG repeats (Hamshere et al. 1997). FISH analysis of DM1 fibroblasts carrying *DMPK* mutant alleles with 50–80 CTG repeats also fails to detect nuclear RNA foci (Amack et al. 1999). However, this failure may be due to relatively low *DMPK* expression levels since RNA foci are detectable following CMV-driven expression of a DMPK 3'-UTR-$(CTG)_{57}$, but not a DMPK 3'-UTR-$(CTG)_{26}$ reporter gene in mouse C2C12 myoblasts (Amack et al. 1999). Because the threshold for nuclear retention is between 26 and 57 repeats, this evidence supports a direct role for the formation of nuclear $(CUG)_n$ foci in DM1 pathogenesis.

Although the $(CCTG)_n$ expansion mutation is located in the first intron of *ZNF9*, these mutant RNAs also appear in nuclear foci in DM2 patient muscle and derived myoblasts (Liquori et al. 2001; Mankodi et al. 2001). Of course, the formation of $(CUG)_n$ and $(CCUG)_n$ RNA foci may be a byproduct, and not the cause, of disease since the number and size of DMPK and ZNF9 RNA foci in patient myofibers do not correlate strictly with disease severity. A problem with this interpretation is that current FISH procedures may not allow the visualization of small RNA foci which may be present, but undetectable, in affected myonuclei.

3.2
$(CUG)_n$ RNA Structures and Gain-of-Function Models

The seminal finding that mutant *DMPK* transcripts accumulated in nuclear foci suggested that these RNAs might be toxic similar to polyQ for the coding region $(CAG)_n$ expansion diseases. How could these RNAs gain a dominant-negative function? Computer modeling indicates that DM1 and DM2 expansion RNAs formed double-stranded (ds)RNA hairpins with U:U mismatches for $(CUG)_n$ and C:U and U:C mismatches for $(CCUG)_n$ (Fig. 2A). DM2 $(CCUG)_n$ RNA hairpins are predicted to be significantly less stable than $(CUG)_n$ repeats which could explain why DM2 disease-associated repeats reach exceptional lengths (Liquori et al. 2001). Experimental confirmation of $(CUG)_n$ dsRNA structure has been achieved using nuclease and lead cleavage susceptibility analysis which indicates that $(CUG)_5$ is predominantly single-stranded, whereas $(CUG)_{>11}$ RNA folds into a "slippery" secondary structure with a metastable stem (Napierala and Krzyzosiak 1997). Nuclease digestion studies have confirmed that longer $(CUG)_n$ RNAs (n = 35, 69 and 140 in this study) fold into single RNA hairpin structures (Tian et al. 2000). The CUG hairpin has been visualized using electron microscopy (EM) of $(CUG)_{130}$ RNAs (Michalowski et al. 1999). Interestingly, $(CAG)_{130}$ folds into a similar hairpin structure that is indistinguishable from $(CUG)_{130}$ at the EM resolution level. Another structural view, based on migration on nondenaturing gels and circular dichroism, is that short $(CUG)_n$ repeats form a linked triangular, or "toblerone", structure and further repeat expansion causes this structure to fold back upon itself (Pinheiro et al. 2002).

Why is the formation of mismatched CUG and CCUG hairpins pathogenic? One possibility is these novel dsRNA structures acquire a function independent of protein activity that is particularly deleterious to muscle cells. Two recent examples of RNA motifs with intrinsic functions include the prokaryotic *RFN* element, which acts as a flavin mononucleotide (FMN)-dependent riboswitch to control translation (Winkler et al. 2002), and the steroid hormone receptor RNA activator (SRA) that serves as a transcriptional coactivator (Lanz et al. 2002). A second possibility is that dsCUG and dsCCUG RNAs might present novel binding sites for nuclear factors that are critical for normal cellular function. According to this hypothesis, mutant RNAs would recruit and sequester these nuclear factors, thus influencing biochemical pathways that require their functions. This sequestration effect might be exacerbated by the inability of affected cells to efficiently degrade dsCUG and dsCCUG which results in the formation of nuclear DM1 and DM2 RNA foci. Therefore, elucidating the role of these nuclear factors in muscle cell function is critical to understanding the molecular etiology of the myotonic dystrophies.

3.3
Guilt by Association: Proteins that Bind CUG and CCUG Repeat Expansions

The hypothesis that *DMPK* mutant allele transcripts are toxic because they sequester RNA binding proteins led to the identification of several CUG-binding proteins (Fig. 3). The first CUG-binding protein (CUGBP1, previously hNab50) characterized was originally identified as a human protein that interacts in the two-hybrid system with yeast Nab2p, a nuclear protein important for both mRNA polyadenylation and nuclear export (Caskey et al. 1996; Green et al. 2002; Hector et al. 2002). Analysis of the RNA-binding properties of CUGBP1 by electrophoretic mobility shift assays (EMSA) revealed that it is a $(CUG)_8$ -binding protein (Timchenko et al. 1996). Two separate EMSA activities have been observed for CUGBP1. While CUG-BP1 activity is predominantly cytoplasmic, CUG-BP2 is nuclear and this latter activity increases about twofold in DM1, compared to normal, lymphoblast nuclear extracts (for this review, we will refer to the protein as CUGBP1 and the EMSA activities as CUG-BP1 and CUG-BP2). An interesting model has been proposed in which CUGBP1 is not only sequestered by *DMPK* mutant allele transcripts, but is also a normal phosphorylation substrate for the DMPK kinase (Roberts et al. 1997). Loss of DMPK activity in DM1 cells leads to accumulation of hypophosphory-

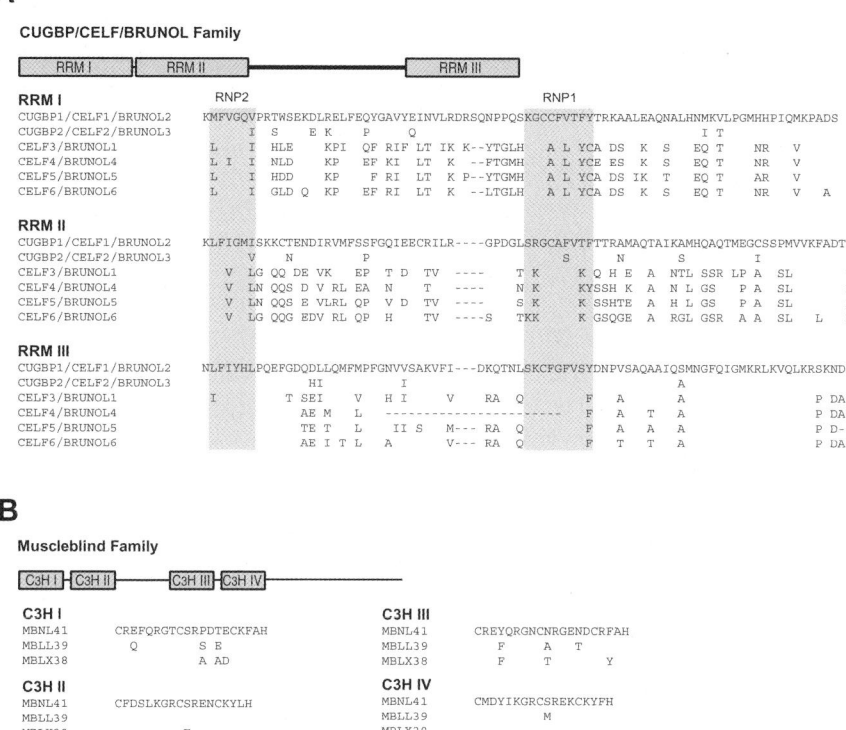

Fig. 3. The CUGBP and muscleblind protein families. **A** Illustration of human CUGBP/CELF/ BRUNOL proteins with three RNA recognition motifs (RRM; top, *boxes*) and the corresponding amino acid sequence shown below (Timchenko et al. 1996; Good et al. 2000; Ladd et al. 2001). The conserved RNP1 and RNP2 motifs in the RRMs are highlighted in the *grey stippled region*. While RRM II and the linker region (*thick line* between RRM II and RRM III) appear to be the most important for binding to UG dinucleotide repeats (Takahashi et al. 2000), RRM III is more conserved in primary sequence between members of this family than either RRM 1 or II. **B** Human muscleblind proteins MBNL41, MBLL39 and MBLX38 are shown with the four CCCH motifs (C_3H, *boxes*). Note that C_3H I and C_3H III motifs (20 amino acids, aa) are longer than C_3H II and C_3H IV (18 aa) and that sequence conservation between the three muscleblind proteins is greater than for the CUGBP/CELF/BRUNOL proteins

lated CUGBP1 (responsible for CUG-BP2 activity) in the nucleus. The RNA-binding properties of these CUGBP1 isoforms have not been studied rigorously, but CUGBP1 appears to be associated with CUG repeat RNA in DM1 heart cytoplasmic extracts, but is predominantly RNA-free in normal heart (Timchenko et al. 2001a).

What is the relationship between the binding of CUGBP1 to CUG repeats and the increasing severity of DM1 disease as the CUG repeat expands? One hypothesis is that CUGBP1 binds directly to mutant DMPK transcripts and is sequestered by these RNAs resulting in a loss of CUGBP1 function. However, several lines of evidence argue against this hypothesis. First, CUG-BP2 activity increases in nuclear extracts from DM1 cells, and CUGBP1 protein levels are elevated in DM1 muscle, compared to normal controls (Timchenko et al. 1996; Roberts et al. 1997; Savkur et al. 2001). The intracellular level of CUGBP1 also increases when $(CUG)_{170}$ and $(CUG)_{500}$ RNAs are expressed in transfected COS7 cells (Timchenko et al. 2001a). In addition, CUGBP1 alternative pre-mRNA splicing activity (see following section) increases in DM1 cells indicating that the protein is not effectively sequestered by $(CUG)_n$ expansions (Philips et al. 1998; Savkur et al. 2001; Charlet-B. et al. 2002; Mankodi et al. 2002). Second, DM1-relevant CUG repeats fold into a stable RNA hairpin as described in the preceding section, but the RNA recognition motifs (RRMs) in CUGBP1 interact primarily with ssRNA (Fig. 3; Varani and Nagai 1998). Indeed, visualization of complexes in the electron microscope (EM) between CUGBP1 and several long CUG repeat RNAs, including $(CUG)_{130}$ and DMPK-3′UTR-CUG$_{90}$, demonstrate that CUGBP1 does not bind stably to either the dsRNA hairpin or the 4–6 nt ssRNA terminal loop (Michalowski et al. 1999; see Fig. 2A). Instead, CUGBP1 associates with ssCUG repeats at the base of the hairpin. Presumably, these ssRNA regions result from out-of-register CUG repeats at both the 5′- and 3′-end. More importantly, CUGBP1 does not colocalize with mutant DMPK transcripts in nuclear RNA foci (Michalowski et al. 1999; Fardaei et al. 2001, 2002). Since DM1 is associated with an increase in the intracellular protein level and nuclear RNA-binding activity of CUG-BP1, expression of large CUG repeats may promote enhanced CUGBP1 protein stability (Timchenko et al. 2001a).

The discovery that CUGBP1 did not colocalize with DM1 nuclear RNA foci enriched in mutant DMPK transcripts suggested that dsCUG-binding proteins might exist which are sequestered by mutant DMPK transcripts. One dsRNA-binding protein that has been implicated in DM1 pathogenesis is the dsRNA-dependent protein kinase PKR since PKR activity is stimulated by dsCUG (Tian et al. 2000). However, recent evidence indicates that PKR activity is not required for the development of the DM1 phenotype since there is no change in the development of myotonia and related muscle structural abnormalities in HSA$_{LR}$ mice when they are crossed with either $Pkr^{+/-}$ or $Pkr^{-/-}$ knockout lines (Mankodi et al., unpubl. data). In contrast, another type of dsRNA-binding protein does appear to be involved in DM1 and DM2. Using HeLa nuclear extracts and a dsCUG photocrosslinking assay, Miller et al. (2000) identified a family of dsCUG-binding proteins, the triplet repeat expansion (EXP) RNA-binding proteins. The EXP proteins in nuclear extracts fail to recognize $(CUG)_{\leq 11}$ RNAs, which are primarily single-stranded, but do cross-link to $(CUG)_{\geq 20}$ dsRNAs and binding to dsCUG is proportional to the length of the RNA hairpin. Remarkably, the EXP proteins are homologues of the *Drosophila* muscleblind proteins required for terminal differentiation of both muscle and photoreceptors, and therefore human EXP proteins were renamed muscleblind or MBNL (Fig. 3; Begemann et al. 1997; Artero et al. 1998; Miller et al. 2000). The human genome database revealed two additional MBNL-related proteins, MBLL (muscleblind-like) and MBLX (muscleblind-like on the X chromosome, also called MBXL and CHCR; Miller et al. 2000; Fardaei et al. 2002; Squillace et al. 2002). In contrast to CUGBP1, all three muscleblind proteins accumulate in nuclear foci in DM1 and DM2 cells and colocalize with mutant *DMPK* and *ZNF9* transcripts in skeletal muscle (Miller et al. 2000; Fardaei et al. 2001, 2002; Mankodi et al. 2001). In agreement with the observation that the muscleblind proteins show a binding preference for dsCUG RNAs, the subnuclear distribution of several unrelated dsRNA-binding proteins is not altered in DM1 or DM2 muscle (Mankodi et al., unpubl. data). A model has been presented in which muscleblind proteins are recruited, and then sequestered, by $(CUG)_n$ expansion RNAs leading to loss of muscleblind activity and disease (Miller et al. 2000).

The normal cellular functions of mammalian muscleblind proteins have not been characterized, but at least one MBNL protein isoform, EXP42, is induced during C2C12 mouse myoblast fusion and myotube formation (Miller et al. 2000). Surprisingly, MBLX/MBXL/CHCR transcript and protein levels decline during this period and constitutive expression of this gene inhibits induction of myosin heavy chain (Squillace et al. 2002). Thus, *MBLX/MBXL/CHCR* gene expression might inhibit myofiber formation which would explain why these transcripts are not detectable in human adult skeletal muscle while *MBNL* gene transcripts are most abundant in this tissue (Miller et al. 2000; Fardaei et al. 2002).

3.4
Cellular Functions Affected by $(CUG)_n$ Expansion

What cellular pathways are affected by the $(CTG)_n$ expansion mutation and is it possible to correlate the resulting dysfunction of these pathways to the length of the CTG repeat expansion? It appears that the regulation of alternative pre-mRNA splicing is affected by the production of large CUG and CCUG repeats. The observation that CUG repeats in muscle-specific splicing enhancers (MSEs) are present in the intron downstream of exon 5 (E5) in chicken and human cardiac troponin T (cTNT) led to the discovery that CUGBP1 promoted E5 inclusion in cTNT mRNA during alternative pre-mRNA splicing (Philips et al. 1998). DM1 patient heart and skeletal muscles show enhanced E5 inclusion activity which is dependent on the intronic MSE CUG repeats, and UV crosslinking studies demonstrate that CUGBP1 binds directly to these repeats. Importantly, E5 inclusion increases when human cTNT is expressed following co-transfection of skeletal muscle cell lines with cTNT and DM1 minigenes, and the level of E5 inclusion is dependent on the length of the DM1 minigene

CTG repeat (Philips et al. 1998). These results suggest that DM1 disease-associated muscle defects result from abnormalities in alternative pre-mRNA splicing.

A critical role for CUGBP1 in alternative splicing regulation in muscle cells is supported by several notable studies. Insulin resistance is a clinical feature associated with DM1 disease due to altered splicing of the heterotetrameric insulin receptor (IR) which is composed of two α-, and two β-, subunits encoded by the *INSR* gene. In DM1 muscle, the IR-A isoform, which shows lower insulin signaling activity than IR-B, predominates due to exclusion of the α-subunit exon 11 (E11) during pre-mRNA splicing (Savkur et al. 2001). Enhancement of E11 exclusion was also seen when NIH 3T3 cells were cotransfected with an IR minigene, containing the region between exons 10 and 12, and either a DMPK minigene carrying 1440 CTG repeats or minigenes expressing CUGBP1. CUGBP1 binds strongly to UG dinucleotide repeats in the yeast three-hybrid system and the intronic enhancer upstream of the α-subunit E11 contains UG repeats (Takahashi et al. 2000; Charlet-B. et al. 2002).

The muscle hyperexcitability characteristic of DM1 and DM2 individuals also appears to be the result of abnormal pre-mRNA splicing. Loss of the major chloride channel in skeletal muscle, ClC-1, is responsible for myotonia in mice (Chen et al. 1997). Recent studies on HSA_{LR} mice indicate that aberrant ClC-1 pre-mRNA splicing leads to elevated levels of abnormal ClC-1 mRNAs containing premature termination codons, which are probably degraded via nonsense-mediated decay (Mankodi et al. 2002). The resulting loss of ClC-1 protein from myofibers leads to defective transmembrane chloride conductance. Related ClC-1 splicing defects have also been uncovered in DM1 and DM2 skeletal muscle, and elevated CUGBP1 levels correspond to these abnormal splicing events (Charlet-B. et al. 2002; Mankodi et al. 2002). Co-transfection of normal QT35 fibroblasts with ClC-1 intron 2 and CUGBP1 minigenes results in enhanced ClC-1 intron 2 inclusion, which is also observed in DM1 and DM2 muscle (Charlet-B. et al. 2002). Similar to the insulin receptor, this ClC-1 intron possesses a UG-rich motif upstream of the 3′ splice site and CUGBP1 interacts specifically with this region.

In summary, alternative splice site selection is affected in myotonic dystrophy. The myotonia associated with both DM1 and DM2 probably results from the loss of correctly processed ClC-1 mRNA. The CUGBP1 protein is an alternative splicing factor that recognizes UG-rich intronic splicing elements to either enhance (cTNT, ClC-1) or repress (IR-A) inclusion of alternatively spliced exons. In DM1 skeletal muscle, CUGBP1 levels are elevated and aberrant splicing of cTNT, IR-A and ClC-1 pre-mRNAs correlates with increased CUGBP1 activity. The relative levels of CUGBP1 in HSA_{SR} versus HSA_{LR} transgenic muscle have not been reported. It is noteworthy that CUGBP1 may also function in additional pathways related to DM1 pathogenesis as well as muscle formation and maintenance. For example, CUGBP1 may be a component of the machinery required for the regulation of muscle differentiation by promoting translation of the cell cycle inhibitor p21 (Timchenko et al. 2001b).

3.5
All Together Now: Loss of Myogenic Differentiation in Congenital DM1 Requires (CUG)$_n$ Expansion and the *DMPK* 3′-UTR

Congenital DM1 (CDM) is the most severe form of myotonic dystrophy and is characterized by severe muscle deficiencies, including hypotonia and associated myofiber immaturity with increased numbers of satellite cells, suggesting that terminal differentiation and maturation of skeletal muscle is compromised (Harper 2001). In cell culture, human quadriceps CDM satellite cells show nuclear (CUG)$_n$ RNA foci, reduced proliferative potential and defective myoblast fusion properties (Furling et al. 2001). Since a congenital form of DM2 has not been documented, loss of myogenic differentiation capacity may require (CTG)$_n$ expansion in the context of *DMPK* allele expression. Specifically, the more severe CDM disorder could be due to the larger CTG expansions seen in this form of the disease and the corresponding increase in *DM1* locus methylation (Sabouri et al. 1993; Steinbach et al. 1998). One intriguing proposal is that increased *DMPK* transcription in CDM is due to reduced interaction of the insulator-binding protein CTCF to two sites that flank the *DMPK* CTG repeat resulting in productive interactions between the enhancer of the downstream *SIX5* gene and the *DMPK* promoter (Filippova et al. 2001).

Differentiation of C2C12 mouse myoblasts is also inhibited by overexpression of either a full-length DMPK cDNA or the DMPK 3′-UTR alone fused to a reporter gene (Sabourin et al. 1997). This 3′-UTR myoblast fusion inhibitory element mapped to a 239 bp region 5′ of the CTG repeat region and loss of fusion activity correlates with a fourfold reduction in myogenin mRNA levels. Other studies have confirmed that myoblast differentiation is blocked by the expression of reporter genes containing the DMPK 3′-UTR, but have concluded that expansion of the CTG repeat region is critical for inhibitory activity (Amack et al. 1999; Bhagwati et al. 1999; Amack and Mahadevan 2001). In support of this conclusion, revertant DMPK 3′-UTR-(CUG)$_{200}$ C2C12 clones that fuse efficiently have specifically deleted the DMPK (CTG)$_n$ expansion and downstream sequences (Amack et al. 1999). Further work confirmed that an expanded CUG repeat tract alone is necessary, but not sufficient, to inhibit C2C12 myogenic differentiation. The implication is that the activity of nuclear factors that bind to the (CUG)$_n$ expansion (muscleblind, CUGBP1) as well as downstream DMPK sequences (hnRNPs C and I/PTB, U2AF, PSF) are affected by nuclear retention of mutant DMPK RNAs (Amack and Mahadevan 2001). What myogenic factors are affected by mutant DMPK 3′-UTR expression? A recent study reports that MyoD protein levels are reduced in C2C12 myoblasts expressing a GFP-DMPK 3′-UTR-(CUG)$_{200}$, but not a GFP-DMPK 3′-UTR-(CUG)$_5$, transgene. Myoblast differentiation is restored by infection with a retrovirus expressing MyoD (Amack et al. 2002).

In summary, the myotonic dystrophies have provided an equal opportunity resource for investigators interested in chromatin structure and stability, transcriptional regulation, RNA processing and nuclear export, mRNA translation/

turnover and protein function and stability. DM1 cardiac conduction, and some skeletal muscle defects result from loss of *DMPK* expression, the myotonia characteristic of DM1 and DM2 can be accounted for by aberrant ClC-1 pre-mRNA splicing with loss of the ClC-1 chloride channel while hypotonia in CDM1 appears to be linked to inhibition of MyoD function during myogenic differentiation. Proteins implicated in these processes include factors that bind to ssCUG (CUGBP/CELF/BRUNOL protein family), dsCUG (MBNL, MBLL, MBLX, PKR) and the DMPK 3′-UTR (hnRNP C, hnRNP I/PTB, U2AF, PSF). Outstanding mechanistic questions include how do $(CUG)_n$ expansion RNAs alter CUGBP1 protein levels and/or activity, how does expression of mutant DMPK 3′-UTR RNA alter MyoD protein levels, do DM1 and DM2 share a common molecular etiology such as loss of muscleblind activity induced by dsCUG expression, are protein and/or RNA turnover machineries inhibited by toxic RNA expression? Clarifying the molecular basis of RNA-mediated pathogenesis for the myotonic dystrophies should accelerate our understanding of other diseases in which toxic RNAs have been implicated, including the spinocerebellar ataxias types 8 (Koob et al. 1999), 10 (Matsuura et al. 2000), 12 (Holmes et al. 2001b) and Huntington's disease-like type 2 (Holmes et al. 2001a; Table 1, reviewed in Ranum and Day 2002).

4
Unstable Microsatellites, Toxic RNA and Implications for Nuclear Structure and Function

Investigations on unstable microsatellites have provided significant contributions to our understanding of nuclear RNA processing and mRNA transport. For the coding region expansion diseases, several genes encode proteins implicated in RNA processing, nuclear export and translation including *SCA1*, *SCA2*, *PABPN1* (Calado et al. 2000; Fernandez-Funez et al. 2000; Kozlov et al. 2001). Molecular analysis of the Fragile X protein FMRP has led to novel insights into axonal mRNP transport and translational regulation in neurons (O'Donnell and Warren 2002). Attempts to test the toxic RNA hypothesis for myotonic dystrophy resulted in the discovery of new families of alternative splicing factors, the CUGBP/CELF/BRUNOL proteins, and dsRNA-binding proteins, the muscleblind family (Timchenko et al. 1996; Good et al. 2000; Miller et al. 2000; Ladd et al. 2001). Will the discovery of toxic RNA foci in DM1 and DM2 provide new details on nuclear structure? Because these foci do not colocalize with characterized subnuclear structures, such as the nucleolus, the peri-nucleolar compartment (PNC), Cajal bodies or PML bodies (Taneja et al. 1995; Mankodi et al. unpubl. data), they may be randomly deposited aggregates of RNA-protein complexes that have overwhelmed the cellular machineries responsible for RNA and protein turnover. Alternatively, RNA foci could localize to uncharacterized subnuclear structures, perhaps enriched in the muscleblind proteins that have not been detected previously by optical or electron microscopy due to their small size or amorphous structure. Future studies should reveal if such structures exist and, if they do, what role these RNA foci play in nuclear function.

References

Amack JD, Mahadevan MS (2001) The myotonic dystrophy expanded CUG repeat tract is necessary but not sufficient to disrupt C2C12 myoblast differentiation. Hum Mol Genet 10:1879–1887

Amack JD, Paguio AP, Mahadevan MS (1999) *Cis* and *trans* effects of the myotonic dystrophy (DM) mutation in a cell culture model. Hum Mol Genet 8:1975–1984

Amack JD, Reagan SR, Mahadevan MS (2002) Mutant DMPK 3′-UTR transcripts disrupt C2C12 myogenic differentiation by compromising MyoD. J Cell Biol 159:419–429

Artero R, Prokop A, Paricio N, Begemann G, Pueyo I, Mlodzik M, Perez-Alonso M, Maylies MK (1998) The *muscleblind* gene participates in the organization of Z-bands and epidermal attachments of *Drosophila* muscles and is regulated by *Dmef2*. Dev Biol 195:131–143

Becher MW, Kotzuk JA, Davis LE, Bear DG (2000) Intranuclear inclusions in oculopharyngeal muscular dystrophy contain poly(A) binding protein 2. Ann Neurol 48:812–815

Begemann G, Paricio N, Kiss I, Perez-Alonso M, Mlodzik M (1997) *muscleblind*, a gene required for photoreceptor differentiation in Drosophila, encodes novel Cys$_3$His-type zinc-finger-containing proteins. Development 124:4321–4331

Bence NF, Sampat RM, Kopito RR (2001) Impairment of the ubiquitin-proteasome system by protein aggregation. Science 292:1552–1555

Berul CI, Maguire CT, Aronovitz MJ, Greenwood J, Miller C, Gehrmann J, Housman D, Mendelsohn ME, Reddy S (1999) DMPK dosage alterations in atrioventricular conduction abnormalities in a mouse myotonic dystrophy model. J Clin Invest 103:R1–R7

Berul CI, Maguire CT, Gehrmann J, Reddy S (2000) Progressive atrioventricular block in a mouse myotonic dystrophy model. J Interven Cardiac Electrophysiol 4:351–358

Bhagwati S, Shafiq SA, Xu W (1999) (CTG)$_n$ repeats markedly inhibit differentiation of the C2C12 myoblast cell line: implications for congenital myotonic dystrophy. Biochim Biophys Acta 1453:221–229

Bowater RP, Wells RD (2001) The intrinsically unstable life of DNA triplet repeats associated with human hereditary disorders. Prog Nucl Acid Res Mol Biol 66:159–202

Brais B, Bouchard J-P, Xie Y-G, Rochefort DL, Chrétien N, Tomé FMS, Lafrenière RG, Rommens JM, Uyama E, Nohira O, Blumen S, Korcyn AD, Heutink P, Mathieu J, Duranceau A, Codère F, Fardeau M, Rouleau GA (1998) Short GCG expansions in the PABP2 gene cause oculopharyngeal muscular dystrophy. Nat Genet 18:164–167

Calado A, Tomé FMS, Brais B, Rouleau GA, Kühn U, Wahle E, Carmo-Fonseca M (2000) Nuclear inclusions in oculopharyngeal muscular dystrophy consist of poly(A) binding protein 2 aggregates which sequester poly(A) RNA. Hum Mol Genet 9:2321–2328

Caskey CT, Swanson MS, Timchenko LT (1996) Myotonic dystrophy: discussion of molecular mechanism. Cold Spring Harbor Symp Quant Biol 61:607–614

Charlet-B. N, Savkur RS, Singh G, Philips AV, Grice EA, Cooper TA (2002) Loss of the muscle-specific chloride channel in type 1 myotonic dystrophy due to misregulated alternative splicing. Mol Cell 10:45–53

Chen M-F, Niggeweg R, Iaizzo PA, Lehmann-Horn F, Jockusch H (1997) Chloride conductance in mouse models is subject to post-transcriptional compensation of the functional Cl⁻ channel 1 gene dosage. J Physiol (Lond) 504:75–81

Davis BM, McCurrach ME, Taneja KL, Singer RH, Housman DE (1997) Expansion of a CUG trinucleotide repeat in the 3′ untranslated region of myotonic dystrophy protein kinase transcripts results in nuclear retention of transcripts. Proc Natl Acad Sci USA 94:7388–7393

Fardaei M, Larkin K, Brook JD, Hamshere MG (2001) In vivo co-localization of MBNL protein with *DMPK* expanded-repeat transcripts. Nucleic Acids Res 29:2766–2771

Fardaei M, Rogers MT, Thorpe HM, Larkin K, Hamshere MG, Harper PS, Brook JD (2002) Three proteins, MBNL, MBLL and MBXL, co-localize in vivo with nuclear foci of expanded-repeat transcripts in DM1 and DM2 cells. Hum Mol Genet 11:805–814

Fernandez-Funez P, Nino-Rosales ML, de Gouyon B, She W-C, Luchak JM, Martinez P, Tuiegano E, Benito J, Capovilla M, Skinner PJ, McCall A, Canal I, Orr HT, Zoghbi HY, Botas J (2000) Identification of genes that modify ataxin-1-induced neurodegeneration. Nature 408:101–106

Filippova GN, Thienes CP, Penn BH, Cho DH, Hu YJ, Moore JM, Klesert TR, Lobanenkov VV, Tapscott SJ (2001) CTCF-binding sites flank CTG/CAG repeats and form a methylation-sensitive insulator at the *DM1* locus. Nat Genet 28:335–343

Freiman RN, Tjian R (2002) A glutamine-rich trail leads to transcription factors. Science 296:2149–2150

Furling D, Coiffier L, Mouly V, Barbet JP, Lacau St Guily J, Taneja K, Gourdon G, Junien C, Butler-Browne GS (2001) Defective satellite cells in congenital myotonic dystrophy. Hum Mol Genet 10:2079–2087

Good PJ, Chen Q, Warner SJ, Herring DC (2000) A family of human RNA-binding proteins related to the *Drosophila* bruno translational regulator. J Biol Chem 275:28583–28592

Green DM, Marfatia KA, Crafton EB, Zhang X, Cheng X, Corbett AH (2002) Nab2p is required for poly(A) RNA export in *Saccharomyces cerevisiae* and is regulated by arginine methylation via Hmt1p. J Biol Chem 277:7752–7760

Hamshere MG, Newman EE, Alwazzan M, Athwal BS, Brook JD (1997) Transcriptional abnormality in myotonic dystrophy affects *DMPK* but not neighboring genes. Proc Natl Acad Sci USA 94:7394–7399

Harper PS (2001) Myotonic dystrophy. Saunders, London

Hector RE, Nykamp KR, Dheur S, Anderson JT, Non PJ, Urbinati CR, Wilson SM, Minvielle-Sebastia L, Swanson MS (2002) Dual requirement for yeast hnRNP Nab2p in mRNA poly(A) tail length control and nuclear export. EMBO J 21:1800–1810

Holmes SE, O'Hearn E, Rosenblatt A, Callahan C, Hwang HS, Ingersoll-Ashworth RG, Fleisher A, Stevanin G, Brice A, Potter NT, Ross CA, Margolis RL (2001a) A repeat expansion in the gene encoding junctophilin-3 is associated with Huntington disease-like 2. Nat Genet 29:377–378

Holmes SE, O'Hearn E, Ross CA, Margolis RL (2001b) SCA12: an unusual mutation leads to an unusual spinocerebellar ataxia. Brian Res Bull 56:397–403

Jansen G, Groenen PJTA, Bächner D, Jap PHK, Coerwinkel M, Oerlemans F, van den Broek W, Gohlsch B, Pette D, Plomp JJ, Molenaar PC, Nederhoff MGJ, van Echteld CJA, Dekker M, Berns A, Hameister H, Wieringa B (1996) Abnormal myotonic dystrophy protein kinase levels produce only mild myopathy in mice. Nat Genet 13:316–324

Kim Y-J, Noguchi S, Hayashi YK, Tsukahara T, Shimizu T, Arahata K (2001) The product of the oculopharyngeal muscular dystrophy gene, poly(A)-binding protein 2, interacts with SKIP and stimulates muscle-specific gene expression. Hum Mol Genet 10:1129–1139

Klement PA, Skinner PJ, Kaytor MD, Yi H, Hersch SM, Clark HB, Zoghbi HY, Orr HT (1998) Ataxin-1 nuclear localization and aggregation: role in polyglutamine-induced disease in SCA1 transgenic mice. Cell 95:41–53

Klesert TR, Otten AD, Bird TD, Tapscott SJ (1997) Trinucleotide repeat expansion at the myotonic dystrophy locus reduces expression of *DMAHP*. Nat Genet 16:402–406

Koob MD, Moseley ML, Schut LJ, Benzow KA, Bird TD, Day JW, Ranum LPW (1999) An untranslated CTG expansion causes a novel form of spinocerebellar ataxia (SCA8). Nat Genet 21:379–384

Kozlov G, Trempe J-F, Khaleghpour K, Kahvejian A, Ekiel I, Gehring K (2001) Structure and function of the C-terminal PABC domain of human poly(A)-binding protein. Proc Natl Acad Sci USA 98:4409–4413

Ladd AN, Charlet-B N, Cooper TA (2001) The CELF family of RNA binding proteins is implicated in cell-specific and developmentally regulated alternative splicing. Mol Cell Biol 21:1285–1296

Lander ES, Linton LM, Birren B, Nusbaum C, Zody MC et al (2001) Initial sequencing and analysis of the human genome. Nature 409:860–921

Lanz RB, Razani B, Goldberg AD, O'Malley BW (2002) Distinct RNA motifs are important for coactivation of steroid hormone receptors by steroid RNA activator (SRA). Proc Natl Acad Sci USA 99:16081–16086

Liquori CL, Ricker K, Moseley ML, Jacobson JF, Kress W, Naylor SL, Day JW, Ranum LPW (2001) Myotonic dystrophy type 2 caused by a CCTG expansion in intron 1 of ZNF9. Science 293:864–867

Mankodi A, Thornton CA (2002) Myotonic syndromes. Curr Opin Neurol 15:545–552

Mankodi A, Logigian E, Callahan L, McClain C, White R, Henderson D, Krym M, Thornton CA (2000) Myotonic dystrophy in transgenic mice expressing an expanded CUG repeat. Science 289:1769–1772

Mankodi A, Urbinati CR, Yuan Q-P, Moxley RT, Sansone V, Krym M, Henderson D, Schalling M, Swanson MS, Thornton CA (2001) Muscleblind localizes to nuclear foci of aberrant RNA in myotonic dystrophy types 1 and 2. Hum Mol Genet 10:2165–2170

Mankodi A, Takahashi MP, Jiang H, Beck CL, Bowers WJ, Moxley RT, Cannon SC, Thornton CA (2002) Expanded CUG repeats trigger aberrant splicing of ClC-1 chloride channel pre-mRNA and hyperexcitability of skeletal muscle in myotonic dystrophy. Mol Cell 10:35–44

Margolis RL, Ross CA (2001) Expansion explosion: new clues to the pathogenesis of repeat expansion neurodegenerative diseases. Trends Mol Med 7:479–482

Matsuura T, Yamagata T, Burgess DL, Rasmussen A, Grewal RP, Watase K, Khajavi M, McCall AE, Davis CF, Zu L, Achari M, Pulst SM, Alonso E, Noebels JL, Nelson DL, Zoghbi HY, Ashizawa T (2000) Large expansion of the ATTCT pentanucleotide repeat in spinocerebellar ataxia type 10. Nat Genet 26:191–194

Michalowski S, Miller JW, Urbinati CR, Paliouras M, Swanson MS, Griffith J (1999) Visualization of double-stranded RNAs from the myotonic dystrophy protein kinase gene and interactions with CUG-binding protein. Nucleic Acids Res 27:3534–3542

Miller JW, Urbinati CR, Teng-umnuay P, Stenberg MG, Byrne BJ, Thornton CA, Swanson MS (2000) Recruitment of human muscleblind proteins to (CUG)$_n$ expansions associated with myotonic dystrophy. EMBO J 19:4439–4448

Nakamura K, Jeong S-Y, Uchihara T, Anno M, Nagashima K, Nagashima T, Ikeda S, Tsuji S, Kanazawa I (2001) SCA17, a novel autosomal dominant cerebellar ataxia caused by an expanded polyglutamine in TATA-binding protein. Hum Mol Genet 10:1441–1448

Napierala M, Kryzyzosiak WJ (1997) CUG repeats present in myotonin kinase RNA form metastable "slippery" hairpins. J Biol Chem 272:31079–31085

Narang MA, Waring JD, Sabourin LA, Korneluk RG (2000) Myotonic dystrophy (DM) protein kinase levels in congenital and adult DM patients. Eur J Hum Genet 8:507–512

Nucifora FC, Sasaki M, Peters MF, Huang H, Cooper JK, Yamada M, Takahashi H, Tsuji S, Troncoso J, Dawson VL, Dawson TM, Ross CA (2001) Interference by huntingtin and atrophin-1 with CBP-mediated transcription leading to cellular toxicity. Science 291:2423–2428

O'Donnell WT, Warren ST (2002) A decade of molecular studies of Fragile X syndrome. Annu Rev Neurosci 25:315–338

Orr HT (2001) Beyond the Qs in the polyglutamine diseases. Genes Dev 15:925–932

Philips AV, Timchenko LT, Cooper TA (1998) Disruption of splicing regulated by a CUG-binding protein in myotonic dystrophy. Science 280:737–741

Pinheiro P, Scarlett G, Rodger A, Rodger PM, Murray A, Brown T, Newbury SF, McClellan JA (2002) Structures of CUG repeats: potential implications for human genetic diseases. J Biol Chem 277:35183–35190

Ranum LPW, Day JW (2002) Dominantly inherited, non-coding microsatellite expansion disorders. Curr Opin Genet Dev 12:266–271

Reddy S, Smith DB, Rich MM, Leferovich JM, Reilly P, Davis BM, Tran K, Rayburn H, Bronson R, Cros D, Balice-Gordon RJ, Housman D (1996) Mice lacking the myotonic dystrophy protein kinase develop a late onset progressive myopathy. Nat Genet 13:325–335

Roberts R, Timchenko NA, Miller JW, Reddy S, Caskey CT, Swanson MS, Timchenko LT (1997) Altered phosphorylation and intracellular distribution of a $(CUG)_n$ triplet repeat RNA-binding protein in patients with myotonic dystrophy and in myotonin protein kinase knockout mice. Proc Natl Acad Sci USA 94:13221–13226

Sabouri LA, Mahadevan MS, Narang M, Lee DS, Surh LC, Korneluk RG (1993) Effect of the myotonic dystrophy (DM) mutation on mRNA levels of the DM gene. Nat Genet 4:233–238

Sabourin LA, Tamai K, Narang MA, Korneluk RG (1997) Overexpression of 3′-untranslated region of the myotonic dystrophy kinase cDNA inhibits myoblast differentiation in vitro. J Biol Chem 272:29626–29635

Sarkar PS, Appukuttan B, Han J, Ito Y, Ai C, Tsai W, Chai Y, Stout JT, Reddy S (2000) Heterozygous loss of Six5 in mice is sufficient to cause ocular cataracts. Nat Genet 25:110–109

Savkur RS, Philips AV, Cooper TA (2001) Aberrant regulation of insulin receptor alternative splicing is associated with insulin resistance in myotonic dystrophy. Nature Genet 29:40–47

Seznec H, Agbulut O, Sergeant N, Savouret C, Ghestem A, Tabti N, Willer J-C, Ourth L, Duros C, Brisson E, Fouquet C, Butler-Browne G, Delacourte A, Junien C, Gourdon G (2001) Mice transgenic for the human myotonic dystrophy region with expanded CTG repeats display muscular and brain abnormalities. Hum Mol Genet 10:2717–2726

Squillace RM, Chenault DM, Wang EH (2002) Inhibition of muscle differentiation by the novel muscleblind-related protein CHCR. Dev Biol 250:218–230

Steinbach P, Gläser D, Vogel W, Wolf M, Schwemmle S (1998) The DMPK gene of severely affected myotonic dystrophy patients is hypermethylated proximal to the largely expanded CTG repeat. Am J Hum Genet 62:278–285

Takahashi N, Sasagawa N, Suzuki K, Ishiura S (2000) The CUG-binding protein binds specifically to UG dinucleotide repeats in a yeast three-hybrid system. Biochem Biophys Res Commun 277:518–523

Taneja KL, McCurrach M, Schalling M, Housman D, Singer RH (1995) Foci of trinucleotide repeat transcripts in nuclei of myotonic dystrophy cells and tissues. J Cell Biol 128:995–1002

Tian B, White RJ, Xia T, Welle S, Turner DH, Mathews MB, Thornton CA (2000) Expanded CUG repeat RNAs form hairpins that activate the double-stranded RNA-dependent protein kinase PKR. RNA 6:79–87

Timchenko NA, Cai Z-J, Welm AL, Reddy S, Ashizawa T, Timchenko LT (2001a) RNA CUG repeats sequester CUGBP1 and alter protein levels and activity of CUGBP1. J Biol Chem 276:7820–7826

Timchenko NA, Iakova P, Cai Z-J, Smith JR, Timchenko LT (2001b) Molecular basis for impaired muscle differentiation in myotonic dystrophy. Mol Cell Biol 21:6927–6938

Timchenko LT, Miller JW, Timchenko NA, DeVore DR, Datar KV, Lin L, Roberts R, Caskey CT, Swanson MS (1996) Identification of a (CUG)n triplet repeat RNA-binding protein and its expression in myotonic dystrophy. Nucleic Acids Res 24:4407–4414

Tiscornia G, Mahadevan M (2000) Myotonic dystrophy: the role of the CUG triplet repeats in splicing of a novel DMPK exon and altered cytoplasmic DMPK mRNA isoform ratios. Mol Cell 5:959–967

Tóth G, Gáspári Z, Jurka J (2000) Microsatellites in different eukaryotic genomes: survey and analysis. Genome Res 10:967–981

Uyama E, Tsukahara T, Goto K, Kurano Y, Ogawa M, Kim Y-J, Uchino M, Arahata K (2000) Nuclear accumulation of expanded PABP2 gene product in oculopharyngeal muscular dystrophy. Muscle Nerve 23:1549–1554

Van den Broek WJAA, Nelen MR, Wansink DG, Coerwinkel MM, Riele H, Groenen PJTA, Wieringa B (2002) Somatic expansion behavior of the $(CTG)_n$ repeat in myotonic dystrophy knock-in mice is differentially affected by Msh3 and Msh6 mismatch-repair proteins. Hum Mol Genet 11:191–198

Varani G, Nagai K (1998) RNA recognition by RNP proteins during RNA processing. Annu Rev Biophys Biomol Struct 27:407–445

Waelter S, Boeddrich A, Lurz R, Scherzinger E, Lueder G, Lehrach H, Wanker EE (2001) Accumulation of mutant huntingtin fragments in aggresome-like inclusion bodies as a result of insufficient protein degradation. Mol Biol Cell 12:1393–1407

Wang J, Pegoraro E, Menegazzo E, Gennarelli M, Hoop RC, Angelini C, Hoffman EP (1995) Myotonic dystrophy: evidence for a possible dominant-negative RNA mutation. Hum Mol Genet 4:599–606

Winkler WC, Cohen-Chalamish S, Breaker RR (2002) An mRNA structure that controls gene expression by binding FMN. Proc Natl Acad Sci USA 99:15908–15913

Zuker M, Mathews DH, Turner DH (1999) Algorithms and thermodynamics for RNA secondary structure prediction: a practical guide. In: Barciszewski J, Clark BFC (eds) RNA biochemistry and biotechnology. NATO ASI series. Kluwer Academic Publ, Dortrecht, pp 11–43

Assembly and Traffic of Small Nuclear RNPs

Edouard Bertrand[1] and Rémy Bordonné[1]

1
Introduction

Eukaryotic cells contain many types of ribonucleoproteins (RNPs) which are complexes formed by RNA molecules associated with proteins. The majority of these RNPs are in the nucleus and can be classified in two groups, the small nuclear RNPs (snRNPs) that function in the maturation of messenger RNAs, and the small nucleolar RNPs that reside in the cell nucleolus and are required for maturation of ribosomal RNAs. Although both types of RNPs function in two different fundamental processes, progress made during the last few years shows that their assembly requires similar protein components. This review focuses on the factors and mechanisms governing snRNPs and snoRNPs bio-genesis as well as on the mechanisms implicated in the sorting of these RNPs to their intracellular destinations.

2
Biogenesis of Small Nuclear RNPs

2.1
U1, U2, U4 and U5 snRNPs

Splicing of nuclear pre-mRNAs in yeast and mammals occurs in a multicom-ponent complex called the spliceosome. The formation of this highly dynamic structure involves the ordered assembly of the U1, U2 and the U4/U6.U5 snRNPs on the pre-mRNA substrate, together with a multitude of non-snRNPs associated protein factors (Burge et al. 1999). In both systems, except for the U6 snRNP, each particle contains a set of common proteins also called Sm proteins (B/B' in mammals, D1, D2, D3, E, F and G), which assemble around the Sm site of the snRNA (Fig. 1; Will and Lührmann 2001). In addition to the common Sm core proteins, the yeast and metazoan U6, U4/U6 and U4/U6.U5 snRNPs also contain seven distinct proteins called Lsm (like Sm) exhibiting

[1]Institut de Génétique Moléculaire de Montpellier, CNRS UMR 5535-IFR 122, 1919 route de Mende, 34293, Montpellier Cedex 5, France. Tel. +33 4 67 61 36 47; Fax. +33 4 67 04 02 31, e-mail: bertrand@igm.cnrs-mop.fr

Progress in Molecular and Subcellular Biology
P. Jeanteur (Ed.): RNA Trafficking and Nuclear Structure Dynamics
© Springer-Verlag Berlin Heidelberg 2003

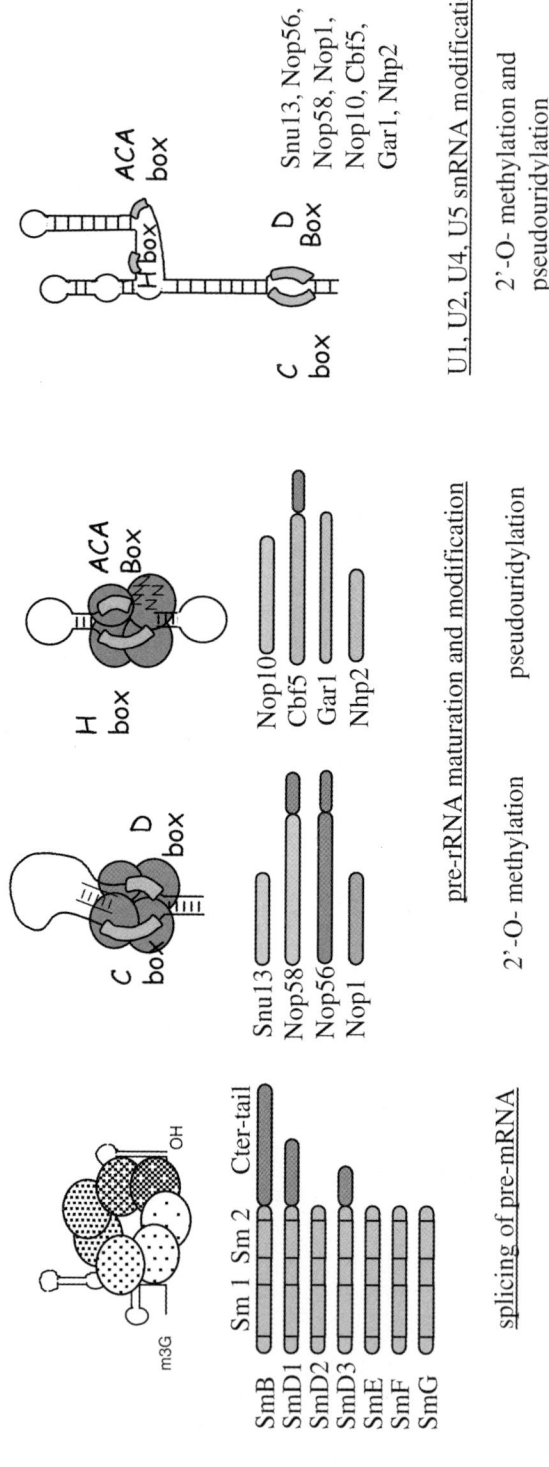

Fig. 1. Structure and function of the different classes of small nuclear RNPs

clear homology to the Sm proteins (Hermann et al. 1995; Seraphin 1995; Gottschalk et al. 1998; Achsel et al. 1999; Mayes et al. 1999; Salgado-Garrido et al. 1999).

The Sm and Lsm proteins are highly conserved in all eukaryotic organisms. They contain the Sm domain, which consists of two conserved regions separated by a linker of variable length (Fig. 1; Hermann et al. 1995; Seraphin 1995). The Sm motif is important for Sm protein function. Indeed, mutations at various positions in this motif, and especially the hydrophobic conserved residues, abolish protein-protein interactions between Sm partners and hinder RNP complex formation (Camasses et al. 1998; Hermann et al. 1995). Structural data show that the Sm domain mediates protein-protein interactions in a heptameric Sm protein doughnut-like structure (Kambach et al. 1999). Multiple protein-RNA contacts occur on the positively charged inner surface of the proposed Sm ring, the most efficient being observed for the SmG an SmB/B' proteins (Urlaub et al. 2001). Interestingly, the architecture of the Sm core domain and RNA binding properties have been conserved during evolution since putative Sm proteins in Archaea form a seven-membered ring, with the RNA interacting inside the central cavity on one face of the doughnut-shaped complex (Toro et al. 2001, 2002).

In higher eukaryotes, assembly of the U-snRNPs is a multistep process requiring nuclear export and import events (Fig. 2). After transcription by RNA polymerase II, the U1, U2, U4 and U5 snRNAs are exported to the cytoplasm. This step requires binding of the snRNAs m^7G cap to the nuclear cap binding complex (CBC) and export to the cytoplasm depends on the CRM1/Xpo1 export receptor (Izaurralde et al. 1995; Fornerod et al. 1997; Stade et al. 1997). This process requires also the PHAX protein which allows formation of snRNA export complexes by mediating the interaction between the CBC/snRNA pre-complex and CRM1 (Ohno et al. 2000).

The Sm proteins, which are stored in the cytoplasm (Mattaj 1986), then assemble onto the snRNA Sm site. In vertebrates, this step requires the survival of motor neurons protein (SMN), the spinal muscular atrophy (SMA) gene product (for reviews, see Frugier et al. 2002; Meister et al. 2002; Paushkin et al. 2002). SMN binds directly to the symmetrical dimethylarginine RG-rich domains of SmB, SmD1 and SmD3 (Brahms et al. 2000, 2001; Friesen and Dreyfuss 2000; Friesen et al. 2001a), and these modifications are formed by a complex containing the methyltransferase JBP1/PRMT5 (Friesen et al. 2001b; Meister et al. 2001).

The association of the complete Sm core on the snRNA is necessary for both 3'-end maturation and for the hypermethylation of the m^7G cap of the snRNA to a methyl-2,2,7-guanosine cap structure (m_3G; Mattaj 1986; Plessel et al. 1994; Raker et al. 1996). In yeast, 3'-extended precursors are stabilized by binding of the La protein and further matured by RNAse III cleavage followed by 3' trimming by the exosome (Chanfreau et al. 1997; Allmang et al. 1999; Seipelt et al. 1999; Xue et al. 2000). The trimethylguanosine synthase responsible for m_3G cap formation has been identified recently in yeast (Mouaikel et al. 2002).

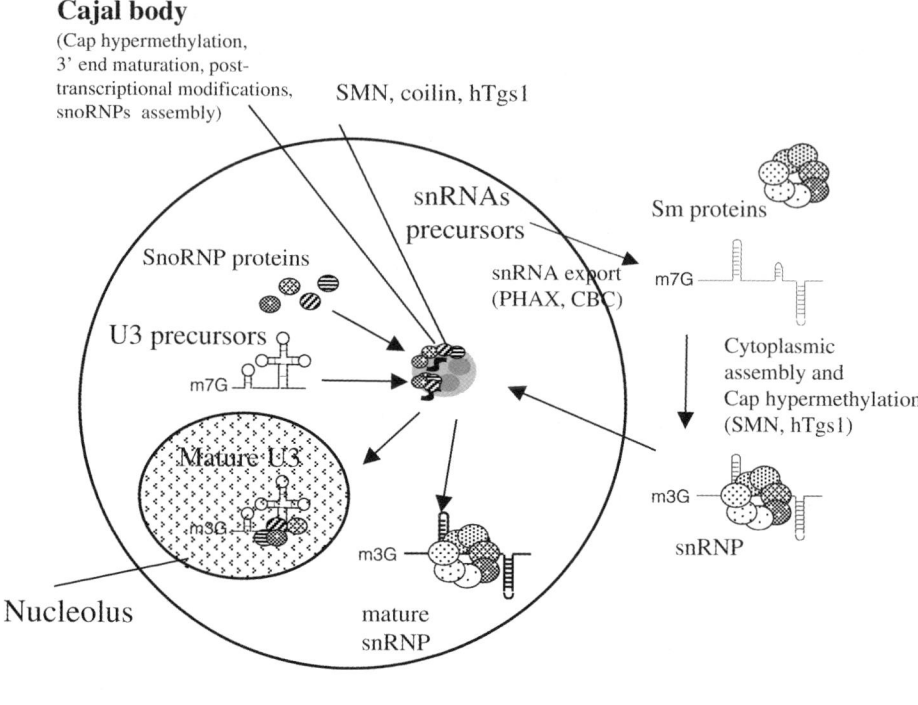

Cajal body
(Cap hypermethylation,
3' end maturation, post-
transcriptional modifications,
snoRNPs assembly)

SMN, coilin, hTgs1

snRNAs precursors

Sm proteins

SnoRNP proteins

snRNA export (PHAX, CBC)

m7G

U3 precursors

m7G

Cytoplasmic assembly and Cap hypermethylation (SMN, hTgs1)

Mature U3

m3G

m3G

snRNP

Nucleolus

m3G

mature snRNP

Human nucleus

Fig. 2. Schematic representation of snRNPs and snoRNPs biogenesis in mammalian cells

This enzyme, named Tgs1, binds preferentially to the C-terminal tail of the yeast SmB protein showing that the yeast enzyme has similar properties to the human hypermethylase which recognizes the U1 snRNP by binding to the SmB/B' proteins (Plessel et al. 1994). Consistent with this, we have recently shown that the human hypermethylase binds strongly to the human SmB protein (Mouaikel et al. 2003). The human hypermethylase also interacts both in vitro and in vivo with the SMN protein suggesting a role of the SMN protein in the recruitment of the hypermethylase and downstream snRNP assembly activities (Mouaikel et al. 2003). Accordingly, it has been shown that the SMN complex is associated with cytoplasmic m^7G-capped snRNPs as well as m_3G-capped particles (Massenet et al. 2002; Narayanan et al. 2002). A role for SMN in snRNA cap hypermethylation is also supported by the fact that the dominant negative mutant, SMNΔN27, blocks the snRNP maturation pathway in the cytoplasm at a step preceding cap hypermethylation (Pellizzoni et al. 1998).

Remarkably, the yeast hypermethylase is located in the nucleolus suggesting that in yeast, the snRNAs cycle through this compartment to undergo cap modification (Mouaikel et al. 2002). This indicates that in yeast and in contrast to the situation in higher eukaryotes, snRNPs biogenesis does not require transit of the snRNAs through the cytoplasm. Instead, snRNA may travel through the nucleolus (Fig. 3). The fact that no homologs of SMN, snurportin1 and PHAX are present in yeast further support this notion (Huber et al. 1998; Ohno et al. 2000).

In addition to snRNAs, the yeast hypermethylase is also responsible for hypermethylation of snoRNAs cap structure (Mouaikel et al. 2002). Whether the human hypermethylase is also responsible for snoRNAs m_3G cap formation remains to be formally demonstrated, but this is likely to be the case since the human enzyme locates in nuclear Cajal bodies, in addition to the cytoplasm (Verheggen et al. 2002; Mouaikel et al. 2003, see below). A nuclear role for the hypermethylase is also consistent with microinjection studies in Xenopus oocytes showing that U1 and U2 snRNAs can be hypermethylated in Xenopus oocytes nuclei if bound to the Sm proteins before injection or if sufficient time is allowed for snRNAs to bind endogenous proteins (Terns et al. 1995; Yu et al. 1998).

Both the Sm core complex and the m_3G cap structure of snRNAs provide signals for subsequent import of the newly assembled core U-snRNPs (Mattaj 1986; Fischer and Lührmann 1990; Hamm et al. 1990). These signals will permit the import of the newly made snRNPs to the nucleus. Studies in yeast gave some clues to the nature of the Sm core NLS. Indeed, it has been shown that the C-terminal tails of yeast SmB and SmD1 proteins possess nuclear localization properties and are essential together for growth (Bordonné 2000). These findings suggested that, in the doughnut-like structure formed by the

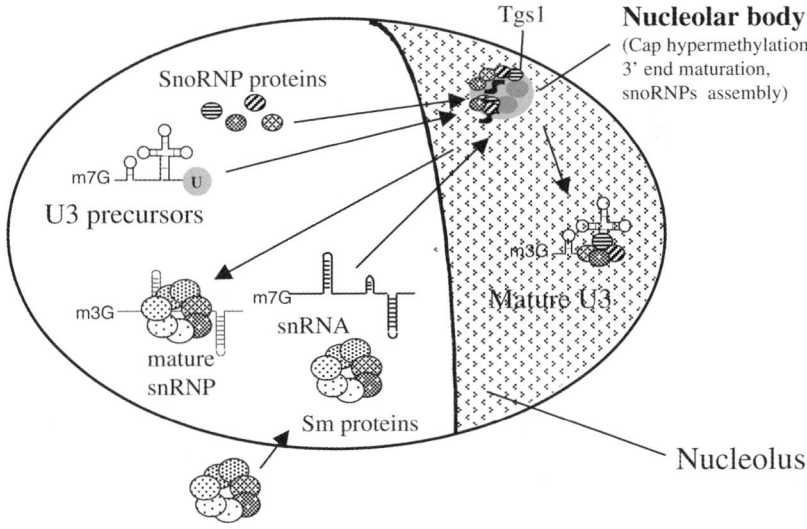

Yeast nucleus

Fig. 3. Schematic representation of snRNPs and snoRNPs biogenesis in yeast

Sm core complex, the carboxyl-terminal extensions of Sm proteins may form a basic amino acid-rich protuberance that functions, at least in part, as a nuclear localization determinant (Bordonné 2000). Given that human SmB, SmD1 and SmD3 also present C-terminal tails rich in basic amino acids, it is likely that the nature of the NLS determinant formed by the human Sm core complex and required for human snRNPs import might also be composed by a protuberance formed by these C-terminal domains.

The import receptor for the m₃G cap structure has been identified in mammals (Huber et al. 1998). This protein, called snurportin1 (SPN1), enhances the m₃G-cap-dependent nuclear import of U snRNPs. SPN1 contains an importin β binding (IBB) domain which allows snRNP to be imported in a Ran- and energy-independent way (Huber et al. 2002). The SPN1 protein recognizes only the m₃G-cap, but not the Sm core NLS indicating that at least two distinct import receptors interact with the snRNP bipartite NLS (Huber et al. 1998). Interestingly, recent studies show that SPN binds preferentially to cytoplasmic SMN complexes and that SMN directly interacts with importin β suggesting that the SMN complex might represent the Sm core NLS receptor (Narayanan et al. 2002).

In addition to the m₃G cap structure, spliceosomal snRNAs also contain many post-transcriptionally modified residues such as 2′-O-ribose-methylated residues and pseudouridines (Reddy and Busch 1988). Modified residues are essential for assembly of functional snRNPs and formation of a productive spliceosome (Segault et al. 1995; Yu et al. 1998). A recent study shows that new guide RNAs called scaRNAs (small Cajal body-specific RNAs) are located specifically in Cajal bodies and mediate internal modifications of the U1, U2, U4 and U5 snRNAs (Darzacq et al. 2002, see below). This suggests that snRNA becomes modified in Cajal bodies (Fig. 2) consistent with the concentration of snRNAs in this compartment following nuclear import (Sleeman and Lamond 1999).

2.2
U6 snRNP

In contrast to the other spliceosomal snRNAs, the biogenesis of the U6 snRNP takes place entirely in the nucleus. After transcription by RNA polymerase III, the U6 snRNA acquires a γ-monomethyl cap structure and the nascent transcripts are stabilized by binding of the La protein to the 3′-end (Xue et al. 2000). Subsequently, the 3′-end of U6 is trimmed to five uridines and converted from a hydroxyl group to a 2′,3′-cyclic phosphate. The La protein is then replaced by the complex of seven Sm-like proteins, Lsm2-Lsm8 (Pannone et al. 1998; Achsel et al. 1999; Mayes et al. 1999; Salgado-Garrido et al. 1999). Binding of the Lsm complex to U6 RNA is also required for subsequent binding of the Prp24 protein to form the mature U6 RNP (Ryan et al. 2002). Posttranscriptional modifications of U6 snRNA occur in the nucleolus which contains all the factors needed for both 2'-O-methylation and pseudouridinylation (Ganot et al. 1999; Tycowski et al. 1998). Consistent with these studies, U6 snRNA localizes transiently to nucleoli after injection into Xenopus oocytes, the 3'-end of U6 RNA being essential and sufficient for this transit, suggesting that it may be mediated by the Lsm proteins (Lange and Gerbi 2000).

2.3
U7 snRNP

In addition to the above described spliceosomal snRNPs, the nuclei of metazoans also contain the U7 snRNP which represents an essential cofactor for 3'-end processing of the replication-dependant histone pre-mRNAs (Dominski and Marzulff 1999). In contrast to U-snRNPs, the Hela U7 snRNP seems particular since it does not have the canonical Sm core composition. Indeed, it lacks the Sm proteins D1 and D2, but contains additional proteins of 14, 50 and 70 kDa (Pillai et al. 2001). Interestingly, the 14-kDa polypeptide was identified as a new Sm-like protein (Lsm10) showing homologies with SmD1 and SmD3. Like the U7 snRNA, Lsm 10 is enriched in nuclear Cajal bodies and its association into U7 snRNPs is dependent on the Sm binding site of the U7 snRNA.

3
Biogenesis of Small Nucleolar RNPs (snoRNPs)

3.1
Principal Features and Functions of snoRNA Families

The first snoRNAs were discovered more than 30 years ago (reviewed in Maxwell and Fournier 1995), and it is now clear that there are more than a hundred stable small RNA residing in human nucleoli (Huttenhofer et al. 2001). It is now also known that some of these RNA are highly conserved across evolution, from Archaebacteria to yeast and human (Gaspin et al. 2000; Omer et al. 2000). In all organisms, the vast majority of these RNAs are predicted to function in the biogenesis of ribosomal RNA (Samarsky and Fournier 1999; Gaspin et al. 2000; Omer et al. 2000; Brown et al. 2001; Huttenhofer et al. 2001). Based on sequence analysis, these RNAs were classified into two families, the box C/D and box H/ACA snoRNAs (Tyc and Steitz 1989; Balakin et al. 1996; Ganot et al. 1997b), with the exception of RNAse MRP which forms a category on its own. The distinctive features of the first snoRNA family are two hexa- and tetra-nucleotides, known as boxes C and D (Tyc and Steitz 1989), which are usually brought together by RNA secondary structure elements to form the box C/D motif (Fig. 1; Maxwell and Fournier 1995). In many cases, box C/D snoRNAs also contain another internal, more divergent box C/D motif, called

C'/D' (Kiss-Laszlo et al. 1998). The second snoRNA family contains other short streches of conserved sequence, box H and box ACA. Similar to box C/D snoRNAs, the general fold of box H/ACA snoRNA is conserved (Fig. 1; Balakin et al. 1996; Ganot et al. 1997b).

While a few snoRNAs, such as U3, were known to be required for the early steps of rRNA processing, namely cleavage of the primary transcript (Kass et al. 1990; Hughes and Ares 1991), the function of the majority of snoRNA was discovered only recently. It was first shown that box C/D snoRNAs act as guides to direct 2'-O-methylation of rRNA bases (Cavaille et al. 1996; Kiss-Laszlo et al. 1996). Indeed, immediately upstream of boxes D and D', these snoRNAs possess 9–21 bases of complementarity with rRNA, and following association with the snoRNA, the 5th hybridized rRNA base becomes methylated (Cavaille et al. 1996; Kiss-Laszlo et al. 1996). Similarly, it was shown later that box H/ACA snoRNAs guide pseudo-uridine formation on rRNAs, also by sequence complementarity (Ganot et al. 1997a; Ni et al. 1997). The length of the hybridized sequence is, however, only 5–6 nucleotides, and the rules determining the modified base are not as clearly defined as for box C/D snoRNAs. Similar to box C/D snoRNA however, each box H/ACA snoRNA can modify two rRNA bases (Ganot et al. 1997a, b; Ni et al. 1997). Remarkably, the enzymes catalyzing the base modifications appear to be stably associated with the snoRNAs. Indeed, fibrillarin (Nop1 in yeast), is a core component of box C/D snoRNP (Schimmang et al. 1989; Tyc and Steitz 1989), and contains signature motifs of S-adenosyl-methionine RNA methyltransferase (Niewmierzycka and Clarke 1999). In addition, point mutations in Nop1 have been shown to block 2'-O-methylation of rRNA in yeast, without affecting snoRNA stability and early rRNA processing (Tollervey et al. 1993). Similarly, dyskerin (cbf5 in yeast), a stable component of box H/ACA snoRNAs, has consensus motifs for pseudo-uridine synthases, and mutations in these motifs block pseudo-uridine formation in yeast (Lafontaine et al. 1998; Watkins et al. 1998; Zebarjadian et al. 1999).

Every snoRNA in each family associates with a small number of core proteins to form a stable RNP. Box C/D snoRNAs bind four proteins: NHPX (Snu13 in yeast, also known as the 15.5 kDa protein), which is an RNA binding protein that directly recognizes the box C/D motif (Watkins et al. 2000); Nop56 and Nop58, two related proteins (Wu et al. 1998; Lafontaine and Tollervey 1999, 2000; Lyman et al. 1999); and fibrillarin (Schimmang et al. 1989; Tyc and Steitz 1989), which binds RNA via a classical RRM domain and is most likely the 2'-O-methylase (Niewmierzycka and Clarke 1999). It was shown recently that assembly of the box C/D snoRNP follows a hierarchical pathway, with NHPX first recognizing the RNA, thereby allowing docking of the other proteins onto the RNA (Watkins et al. 2002). Remarkably, NHPX binds the box C/D motif in a way that leaves most of its surface exposed and that of the RNA (Vidovic et al. 2000). In addition, the bend in the RNA helices induced by NHPX may facilitate the specific recognition of adjacent nucleotides by fibrillarin and Nop58 (Cahill et al. 2002; Watkins et al. 2002). Box H/ACA snoRNP contain

Nhp2, which is related to NHPX and also binds RNA, Nop10, Gar1, and dys-kerin, the pseudo-uridine synthase (Balakin et al. 1996; Henras et al. 1998; Lafontaine et al. 1998; Watkins et al. 1998). Remarkably, box C/D and box H/ACA snoRNP proteins contain a number of related protein motifs. For instance, both fibrillarin and Gar1 possess RG domains, both cbf5 and Nop56/Nop58 contain a basic C-terminus consisting of repeated KKD/E peptides, and Nhp2 and Snu13 are related proteins. This suggest that the two snoRNP fam-ilies may have derived from one another, and/or that they interact with several common factors through conserved interfaces (see below).

3.2
snoRNA Synthesis and Genomic Organization

snoRNA primary transcripts can be synthesized in a variety of ways (for review, see Maxwell and Fournier 1995). First, similar to snRNAs, they can be transcribed by RNA polymerase II as independent units, and in this case the promoter and terminator elements are very similar to that of snRNAs (Stein-metz et al. 2001). An exception to this is the plant U3 gene, which is the sole known snoRNA to be transcribed by RNA polymerase III (Kiss et al. 1991). Second, they can be organized into polycistronic units (Leader et al. 1997), and in this case it is usually an endoribonuclease that processes the primary tran-script into individual units (Chanfreau et al. 1998). Finally, they can be located within introns of mRNAs, and in this case they are processed either following splicing, from the debranched lariat, or by a nuclease that cleaves within the intron (Ooi et al. 1998; Petfalski et al. 1998; Villa et al. 1998). Remarkably, it appears that the primary product of some genes that contain snoRNA in their introns are the snoRNA themselves, because the spliced mRNA contains no discernible open reading frame (Tycowski et al. 1996). It is also interesting that different organisms display a preferential genetic organization. For instance, in yeast and plant snoRNA are often clustered and produced as poly-cistronic units, while in animals, this organization is virtually absent and the vast major-ity of snoRNAs is produced from introns (Brown et al. 2001; Huttenhofer et al. 2001). In animals, intronic snoRNA genes additionally belong to a family of genes known as 5′ TOP, which all contain an oligo-pyrimidine stretch at their transcription start site (Pelczar and Filipowicz 1998; Smith and Steitz 1998). The functional significance of this observation is not clear, but could involve transcriptional and/or translational regulation, as well as a particular intra-cellular trafficking pathway. The genetic organization of snoRNA genes is of some importance because it often determines the nature of the RNA 5′-end. Indeed, intronic and polycistronic snoRNAs usually have naked 5′ and 3′ OH ends that are located a few nucleotides away from the conserved box elements and common structural folds. In contrast, many snoRNAs that are transcribed as independent units retain the 5′-end of the primary transcript including the cap structure. This allows them to carry additional sequences at their 5′-end, which can be essential for their function as in the case of U3 snoRNA. Similar to snRNAs, the cap of these transcripts is hyper-methylated (Reddy and Busch 1988).

3.3
Trafficking of snoRNA and snoRNP Proteins

By definition, snoRNAs are localized in the nucleolus, and initial trafficking studies determined the sequence elements required for this localization. In both yeast and vertebrates, it was shown that the box C/D motif is necessary and sufficient for nucleolar localization (Lange et al. 1998; Samarsky et al. 1998; Narayanan et al. 1999a). Surprisingly, in vertebrates, the box C/D motif was also sufficient to target reporter RNAs to Cajal bodies (Samarsky et al. 1998), a nuclear compartment that is biochemically and spatially related to nucleoli, and was, in fact, known to contain endogenous box C/D snoRNA and snoRNP proteins (Gall 2000). Later, it was shown that this steady-state localization reflected a temporal pathway, and that box C/D snoRNA reporters microinjected into Xenopus oocytes first localized to Cajal bodies, and only later to nucleoli (Narayanan et al. 1999a). Similarly, work with box H/ACA snoRNA showed that boxes H and ACA, when properly folded, are sufficient to target reporter RNA to Cajal bodies and nucleoli, although no clear temporal pathways could be detected in this case (Lange et al. 1999; Narayanan et al. 1999b).

Several studies have shed some light on the role of this transit through Cajal bodies (Verheggen et al. 2002; Mouaikel et al. 2003). Using U3 as a model system, it was shown that U3 precursors are present at the snoRNA transcription site and in Cajal bodies, while the mature RNA is detected in Cajal bodies and nucleoli. Remarkably, these U3 precursors are bound by NHPX, but lack fibrillarin and Nop58, suggesting that snoRNP assembly takes place in Cajal bodies (Fig. 2; Verheggen et al. 2002). This is further confirmed by the fact that these precursors contain a monomethyl cap structure, and that the enzyme likely responsible for cap hypermethylation, Tgs1, localizes in Cajal bodies (Verheggen et al. 2002), and may be docked onto the RNA by associating with hNop56 and hNop58 (Mouaikel et al. 2002). A function of Cajal bodies in snoRNP assembly is also in agreement with the probable role of SMN in this process, which has been shown to associate with both fibrillarin and Gar1 in vivo, to block snoRNP assembly when mutated, and to localize in Cajal bodies (Jones et al. 2001; Pellizzoni et al. 2001).

Recently, the trafficking of several box C/D proteins has been studied in living cells (Leung and Lamond 2002). Remarkably, it has been shown that NHPX follows a very specific trafficking pathway when entering the nucleus: it first localizes to speckles, a compartment involved in mRNA transcription and processing, before localizing in nucleoli (Leung and Lamond 2002). It is also present in Cajal bodies at all times. In addition, NHPX populations localized in speckles and nucleoli do not exchange. This suggests that NHPX associates with intronic snoRNA precursors as it does with U3 precursors, and labels the route taken by the RNA in living cells. It is remarkable that this is a property unique to NHPX, as fibrillarin localizes directly to Cajal bodies and nucleoli, and never to speckles (Leung and Lamond 2002), consistent with the idea that NHPX and fibrillarin are assembled together into box C/D snoRNPs at the level of Cajal bodies.

Yeast cells do not have Cajal bodies, and studies analyzing box C/D snoRNA trafficking suggest that in this case the nucleolus could be involved in snoRNP assembly and maturation (Fig. 3; Verheggen et al. 2002). Indeed, Tgs1, the cap hypermethylase, is concentrated in the nucleolus (Mouaikel et al. 2002), and unassembled U3 precursors can be found in this compartment (Verheggen et al. 2002). Remarkably, under certain growth conditions, it is possible to segregate within the yeast nucleolus the activities devoted to the biogenesis of either rRNA, or sn- and snoRNAs. Indeed, when grown on solid media, Tgs1 localizes in a dot in the nucleolus in about 25% of the cells. This domain was called the nucleolar body, and was also shown to contain U3 precursors, but not the mature form that is mostly excluded from it (Verheggen et al. 2002). These data suggest that the nucleolar activities involved in the biogenesis of small RNAs have some ability to self-assemble and to form a distinct structure, and this may have been at the origin of the evolution of Cajal bodies into an independent compartment. The link between the two structures is further supported by the fact that they concentrate similar markers (Verheggen et al. 2002).

The use of yeast cells to study snoRNA trafficking also allowed to determine the role of individual snoRNP proteins. In the case of box C/D snoRNA, it was surprisingly found that each snoRNP protein was required for RNA localization (Verheggen et al. 2001). This suggested that the determinant for nucleolar localization did not reside on a single protein, but was instead split and dispersed. Possibly, snoRNPs are stabilized within the nucleolus through multiple weak interactions. Alternatively, it is also possible that the snoRNP proteins play different roles that were not resolved by the assay used. For instance, they could play sequential roles in RNA targeting.

4
Noncanonical Types of Box C/D and Box H/ACA RNAs

In addition to canonical snoRNAs, eukaryotic cells contain a number of related RNA that are unlikely to be involved in ribosome biogenesis, because they do not possess any sequence complementarity with rRNA. Instead, some of these RNAs are complementary to snRNAs, at positions known to be modified in vivo (Jady and Kiss 2001). These RNAs are further unusual in that some of them are chimeric and contain canonical motifs for both box C/D and box H/ACA families (Fig. 1), and indeed associate with both sets of core snoRNP proteins (Jady and Kiss 2001). Recently, these RNAs were shown to specifically localize in the Cajal body, where snRNAs also concentrate (Darzacq et al. 2002).

This suggests that these unusual C/D and H/ACA RNAs catalyze base modification of snRNA in Cajal bodies (see above), and they were thus named scaRNAs, for Small Cajal RNAs. These RNA are conserved from Drosophila to humans (Jady and Kiss 2001), and may be absent from yeast, in which case snRNA base modification appears to be catalyzed by protein-only enzymes (Massenet et al. 1999). In addition, it should be noted that modification guides for U6 snRNA do not belong to this family, but are localized in the nucleolus, consistent with the particular trafficking pathway of this RNA (Tycowski et al. 1998; Ganot et al. 1999).

In addition to these RNAs, humans have ubiquitously expressed or tissue-specific snoRNAs that do not contain any sequence complementarity with rRNA or snRNA (Huttenhofer et al. 2001), and their modification guide sequences are, in fact, not generally well conserved between close species (Cavaille et al. 2001). These snoRNAs are thus unlikely to function in ribosome synthesis or snRNA biogenesis, despite the fact that in all cases analyzed to date, they were localized in the nucleolus (Cavaille et al. 2001). Remarkably, a number of these snoRNAs are expressed from imprinted genes, suggesting that they may have entirely novel roles in animals (Cavaille et al. 2000, 2001).

A last related RNA that should be mentioned here is the telomerase RNA. Indeed, telomerase functions as an RNP (Greider and Blackburn 1987), and its RNA moiety provides a binding site and the substrate sequence for the telomerase reverse transcriptase. In higher eukaryotes, the RNA also comply with the general fold and conserved sequences of the box H/ACA snoRNA family (Mitchell et al. 1999). Furthermore, cell fractionation experiments and immuno-precipitation studies have indeed shown that it localizes in nucleoli, and associates with box H/ACA core snoRNP proteins (Mitchell et al. 1999). Remarkably, in yeast, telomerase RNA is not a box H/ACA snoRNA, but rather associates with Sm proteins (Seto et al. 1999). This suggests that either an Sm binding site or a box H/ACA motif can supply the functions required for this RNA, such as stability, localization, and association with various factors. These observations suggest that the two RNA families, sn- and snoRNAs, are functionally related. Telomerase RNA is also particular because its function implies that the RNP should leave the nucleolus to elongate telomeres, demonstrating that despite their nucleolar localization at steady state, snoRNA may carry essential functions elsewhere in the nucleus. This also raises the possibility that the intranuclear localization of snoRNA might be regulated, and recent studies suggest that indeed, telomerase RNA shuttles between the nucleolus and the nucleoplasm (Wong et al. 2002).

5
Similarities Between snRNA and snoRNA Biogenesis

In yeast, sn- and snoRNAs are likely to follow a similar trafficking pathway that is entirely nuclear and involves transit through the nucleolus. RNA processing also requires similar processing factors. For instance, snRNA and a number of snoRNA, such as U3, are transcribed from similar promoters, and transcription is terminated at their 3′-end by similar machineries (Steinmetz et al. 2001). End processing also requires similar factors, the RNAse III ortholog Rnt1 and the exosome at the 3′-end (Chanfreau et al. 1997, 1998; Ooi et al. 1998; Petfalski et al. 1998; Villa et al. 1998), and Tgs1 at the 5′-end (Mouaikel et al. 2002). Furthermore, sn- and snoRNAs precursors are stabilized by the same proteins, including the yeast La homolog (Kufel et al. 2000; Xue et al. 2000).

In vertebrates, sn- and snoRNA also utilize common processing factors, such as SMN and Tgs1, which likely play similar roles in RNP assembly and cap modification for the two RNA families. Tgs1 binds the basic C-terminus of Cfb5, Nop56, Nop58, and SmB, allowing it to access the cap of the RNA (Mouaikel et al. 2002). Similarly, SMN binds dimethylated arginine residues of Sm proteins, and also RG repeats of Gar1 and fibrillarin, thereby enabling the SMN complex to chaperone RNP assembly (Friesen et al. 2001a; Pellizzoni et al. 2001; Narayanan et al. 2002). Sn- and snoRNPs thus appear more closely related than previously appreciated. This is further emphasized by the facts that U4 snRNP and box C/D snoRNPs have one protein in common, NHPX (Watkins et al. 2000), and that telomerase RNA belongs to either family, depending on the species (Mitchell et al. 1999; Seto et al. 1999).

Despite this high level of similarity in their processing pathways, the intracellular trafficking of sn- and snoRNAs is very different in higher eukaryotes: while snRNA precursors are exported to the cytoplasm via PHAX and CRM1 (Ohno et al. 2000), snoRNA precursors appear to be routed to Cajal bodies via an uncharacterized pathway. SMN and Tgs1 thus act in different places, the cytoplasm for snRNAs, and Cajal bodies for snoRNAs. The reason for this is not clear. Because PHAX is absent from yeast, it is possible that snRNA export and cytoplasmic assembly is a recent addition during evolution. Perhaps the very large amount of snRNAs required to process the numerous introns of higher eukaryotes has made cytoplasmic snRNA assembly more beneficial. In particular, this avoids the presence of free Sm proteins in the nucleus, which may be assembled onto inappropriate RNA, or have a deleterious effect on the biogenesis of other nuclear RNA, for instance by titrating nuclear SMN and interfering with its function in snoRNA assembly.

6
snRNAs and snoRNAs: a Common Origin?

As described in the present review, sn- and snoRNA share a number of common features. This may have arisen by convergent evolution, or because sn- and snoRNAs share a common origin. In this respect, it is interesting to note that there is a very ancient link between snoRNAs and introns. Indeed, not only are modern snoRNA often expressed from mRNA introns, but even archaeal tRNA introns contain methylation guide box C/D sRNAs, the ancestors of snoRNAs (Gaspin et al. 2000; Omer et al. 2000). Because snRNA has been suggested to have evolved from autocatalytic, transposable introns (Valadkhan and Manley 2001), this led us to speculate and suggest the following scenario. Possibly, an autocatalytic intron was inserted within a box C/D sRNA. Following mutations at the junctions and evolution of the resulting molecule, the chimeric RNA species may have lost the ability to self-splice, but may have, nevertheless, retained some functional parts of the ribozyme. This could have enabled it to carry splicing in *trans*, the ribozymic moiety providing for catalysis functions, and the sRNA part for stability and RNA processing. This could have been beneficial because such an RNA would have allowed inactivation of transposable introns at other genetic locations, while still excising them in trans. It could have also progressively expanded the repertoire of expressed proteins. Finally, this hypothetical sRNA may have further evolved, and may have lost the canonical boxes C and D, in which case RNA stability could have been provided by proteins bound to box C/D sRNA precursors. In this regard, it is interesting to note that Sm proteins are closely related to, and may have evolved from Lsm proteins, which are known to bind snoRNA precursors in yeast (Kufel et al. 2000).

Acknowledgments. We would like to thank E. Basyuk for critical reading of the manuscript. This work was supported by CNRS, AFM, MNRT (ACI), ECS, and EMBO YIP program.

References

Achsel T, Brahms H, Kastner B, Bacgi A, Wilm M, Lührmann R (1999) A doughnut-shaped heteromer of human Sm-like proteins binds to the 3'-end of U6 snRNA, thereby facilitating U4/U6 formation in vitro. EMBO J 18:5789–5802

Allmang C, Kufel J, Chanfreau G, Mitchell P, Petfalski E, Tollervey D (1999) Functions of the exosome in rRNA, snoRNA and snRNA synthesis. EMBO J 18:5399–5410

Balakin A, Smith L, Fournier M (1996) The RNA world of the nucleolus: two major families of small RNAs defined by different box elements with related functions. Cell 86:823–834

Bordonné R (2000) Functional characterization of nuclear localization signals in yeast Sm proteins. Mol Cell Biol 20:7943–7954

Brahms H, Raymackers J, Union A, de Keyser F, Meheus L, Lührmann R (2000) The C-terminal RG dipeptide repeats of the spliceosomal Sm proteins D1 and D3 contain symmetrical dimethylarginines, which form a major B-cell epitope for anti-Sm autoantibodies. J Biol Chem 275:17122–17129

Brahms H, Meheus L, de Brabandere V, Fischer U, Lührmann R (2001) Symmetrical dimethylation of arginine residues in spliceosomal Sm protein B/B' and the Sm-like protein LSm4, and their interaction with the SMN protein. RNA 7:1531–1542

Brown J, Clark G, Leader D, Simpson C, Lowe T (2001) Multiple snoRNA gene clusters from Arabidopsis. RNA 7:1817–1832

Burge C, Tuschl T, Sharp P (1999) Splicing of precursors to mRNA by the spliceosome. In: Gesteland R, Cech T, Atkins J (eds) The RNA world. Cold Spring Harbor Laboratory Press, Cold Spring Harbor, NY, pp 525–560

Cahill N, Friend K, Speckmann W, Li Z, Terns R, Terns M, Steitz J (2002) Site-specific cross-linking analyses reveal an asymmetric protein distribution for a box C/D snoRNP. EMBO J 21:3816–3828

Camasses A, Bragado-Nilsson E, Martin R, Séraphin B, Bordonné R (1998) Interactions within the yeast Sm core complex: from proteins to amino acids. Mol Cell Biol 18:1956–1966

Cavaille J, Nicoloso M, Bachellerie J (1996) Targeted ribose methylation of RNA in vivo directed by tailored antisense RNA guides. Nature 383:732–735

Cavaille J, Buiting K, Kiefmann M, Lalande M, Brannan C, Horsthemke B, Bachellerie J, Brosius J, Huttenhofer A (2000) Identification of brain-specific and imprinted small nucleolar RNA genes exhibiting an unusual genomic organization. Proc Natl Acad Sci USA 97:14311–14316

Cavaille J, Vitali P, Basyuk E, Huttenhofer A, Bachellerie J (2001) A novel brain-specific box C/D small nucleolar RNA processed from tandemly repeated introns of a noncoding RNA gene in rats. J Biol Chem 276:26374–26383

Chanfreau G, Elela S, Ares M Jr, Guthrie C (1997) Alternative 3'-end processing of U5 snRNA by RNAse III. Genes Dev 11:2741–2751

Chanfreau G, Rotondo G, Legrain P, Jacquier A (1998) Processing of a dicistronic small nucleolar RNA precursor by the RNA endonuclease Rnt1. EMBO J 17:3726–3737

Darzacq X, Jady B, Verheggen C, Kiss A, Bertrand E, Kiss T (2002) Cajal body-specific small nuclear RNAs: a novel class of 2'-O-methylation and pseudouridylation guide RNAs. EMBO J 21:2746–2756

Dominski Z, Marzulff W (1999) Formation of the 3'-end of histone mRNA. Gene 239:1–14

Fischer U, Lührmann R (1990) An essential role for the m_3G cap in the transport of U1 snRNP to the nucleus. Science 249:786–790

Fornerod M, Ohno M, Yoshida M, Mattaj I (1997) CRM1 is an export receptor for leucine-rich nuclear export signals. Cell 90:1051–1060

Friesen W, Dreyfuss G (2000) Specific sequences of the Sm and Sm-like (Lsm) proteins mediate their interaction with the spinal muscular atrophy disease gene product (SMN). J Biol Chem 275:26370–26375

Friesen W, Massenet S, Paushkin S, Wyce A, Dreyfuss G (2001a) SMN, the product of the spinal muscular atrophy gene, binds preferentially to dimethylarginine-containing protein targets. Mol Cell 7:1111–1117

Friesen W, Paushkin S, Wyce A, Massenet S, Pesiridis G, van Duyne G, Rappsilber J, Mann M, Dreyfuss G (2001b) The methylosome, a 20S complex containing JBP1 and pICln, produces dimethylarginine-modified Sm proteins. Mol Cell Biol 21:8289–8300

Frugier T, Nicole S, Cifuentes-Diaz C, Melki J (2002) The molecular bases of spinal muscular atrophy. Curr Opin Genet De 12:294–298

Gall J (2000) Cajal bodies: the first 100 years. Annu Rev Cell Dev Biol 16:273–300

Ganot P, Bortolin M, Kiss T (1997a) Site-specific pseudouridine formation in preribosomal RNA is guided by small nucleolar RNAs. Cell 89:799–809

Ganot P, Caizergues-Ferrer M, Kiss T (1997b) The family of box ACA small nucleolar RNAs is defined by an evolutionarily conserved secondary structure and ubiquitous sequence elements essential for RNA accumulation. Genes Dev 11:941–956

Ganot P, Jady B, Bortolin M, Darzacq X, Kiss T (1999) Nucleolar factors direct the 2'-O-ribose methylation and pseudouridylation of U6 spliceosomal RNA. Mol Cell Biol 19:6906–6917

Gaspin C, Cavaille J, Erauso G, Bachellerie J (2000) Archaeal homologs of eukaryotic methylation guide small nucleolar RNAs: lessons from the Pyrococcus genomes. J Mol Biol 297:895–906

Gottschalk A, Tang J, Puig O, Salgado J, Neubauer G, Colot H, Mann M, Séraphin B, Rosbash M, Lührmann R, Fabrizio P (1998) A comprehensive biochemical and genetic analysis of the yeast U1 snRNP reveals five novel proteins. RNA 4:374–393

Greider C, Blackburn E (1987) The telomere terminal transferase of Tetrahymena is a ribonucleoprotein enzyme with two kinds of primer specificity. Cell 24:887–898

Hamm J, Darzynkiewicz E, Tahara S, Mattaj I (1990) The trimethylguanosine cap structure of U1 snRNA is a component of a bipartite nuclear targeting signal. Cell 62:569–577

Henras A, Henry Y, Bousquet-Antonelli C, Noaillac-Depeyre J, Gelugne J, Caizergues-Ferrer M (1998) Nhp2p and Nop10p are essential for the function of H/ACA snoRNPs. EMBO J 19:7078–7090

Hermann H, Fabrizio P, Raker V, Foulaki K, Hornig H, Brahms H, Lührmann R (1995) snRNP Sm proteins share two evolutionarily conserved sequence motifs which are involved in Sm protein-protein interactions. EMBO J 14:2076–2088

Huber J, Cronshagen U, Kadokura M, Marshallsay C, Wada T, Sekine M, Lührmann R (1998) Snurportin1, an m₃G-cap-specific nuclear import receptor with a novel domain structure. EMBO J 17:4114–4126

Huber J, Dickmanns A, Lührmann R (2002) The importin-beta binding domain of snurportin1 is responsible for the Ran- and energy-independent nuclear import of spliceosomal U snRNPs in vitro. J Cell Biol 156:467–479

Hughes J, Ares M Jr (1991) Depletion of U3 small nucleolar RNA inhibits cleavage in the 5′ external transcribed spacer of yeast pre-ribosomal RNA and impairs formation of 18S ribosomal RNA. EMBO J 10:4231–4239

Huttenhofer A, Kiefmann M, Meier-Ewert S, O'Brien J, Lehrach H, Bachellerie J, Brosius J (2001) RNomics: an experimental approach that identifies 201 candidates for novel, small, non-messenger RNAs in mouse. EMBO J 20:2943–2953

Izaurralde E, Lewis J, Gamberi C, Jarmolowski A, McGuigan C, Mattaj I (1995) A cap-binding protein complex mediating U snRNA export. Nature 376:709–712

Jady B, Kiss T (2001) A small nucleolar guide RNA functions both in 2′-O-ribose methylation and pseudouridylation of the U5 spliceosomal RNA. EMBO J 20:541–551

Jones K, Gorzynski K, Hales C, Fischer U, Badbanchi F, Terns R, Terns M (2001) Direct interaction of the spinal muscular atrophy disease protein SMN with the small nucleolar RNA-associated protein fibrillarin. J Biol Chem 276:38645–38651

Kambach C, Walke S, Young R, Avis J, de la Fortelle E, Raker V, Lührmann R, Li J, Nagai K (1999) Crystal structures of two Sm protein complexes and their implications for the assembly of the spliceosomal snRNPs. Cell 96:375–387

Kass S, Tyc K, Steitz J, Sollner-Webb B (1990) The U3 small nucleolar ribonucleoprotein functions in the first step of preribosomal RNA processing. Cell 60:897–908

Kiss T, Marshallsay C, Filipowicz W (1991) Alteration of the RNA polymerase specificity of U3 snRNA genes during evolution and in vitro. Cell 65:517–526

Kiss-Laszlo Z, Henry Y, Bachellerie J, Caizergues-Ferrer M, Kiss T (1996) Site-specific ribose methylation of preribosomal RNA: a novel function for small nucleolar RNAs. Cell 85:1077–1088

Kiss-Laszlo Z, Henry Y, Kiss T (1998) Sequence and structural elements of methylation guide snoRNAs essential for site-specific ribose methylation of pre-rRNA. EMBO J 17:797–807

Kufel J, Allmang C, Chanfreau G, Petfalski E, Lafontaine D, Tollervey D (2000) Precursors to the U3 small nucleolar RNA lack small nucleolar RNP proteins but are stabilized by La binding. Mol Cell Biol. 20:5415–5424

Lafontaine D, Tollervey D (1999) Nop58p is a common component of the box C+D snoRNPs that is required for snoRNA stability. RNA 5:455–467

Lafontaine D, Tollervey D (2000) Synthesis and assembly of the box C+D small nucleolar RNPs. Mol Cell Biol 20:2650–2659

Lafontaine D, Bousquet-Antonelli C, Henry Y, Caizergues-Ferrer M, Tollervey D (1998) The box H + ACA snoRNAs carry cbf5p, the putative rRNA pseudouridine synthase. Genes Dev 12:527–537

Lange T, Gerbi S (2000) Transient nucleolar localization of U6 small nuclear RNA in *Xenopus laevis* oocytes. Mol Biol Cell 11:2419–2428

Lange T, Borovjagin A, Maxwell E, Gerbi S (1998) Conserved boxes C and D are essential nucleolar localization elements of U14 and U8 snoRNAs. EMBO J 17:3176–3187

Lange T, Ezrokhi M, Amaldi F, Gerbi S (1999) Box H and box ACA are nucleolar localization elements of U17 small nucleolar RNA. Mol Biol Cell 10:3877–3890

Leader D, Clark G, Watters J, Beven A, Shaw P, Brown J (1997) Clusters of multiple different small nucleolar RNA genes in plants are expressed as and processed from polycistronic pre-snoRNAs. EMBO J 16:5742–5751

Leung A, Lamond A (2002) In vivo analysis of NHPX reveals a novel nucleolar localization pathway involving a transient accumulation in splicing speckles. J Cell Biol 157:615–629

Lyman S, Gerace L, Baserga S (1999) Human Nop5/Nop58 is a component common to the box C/D small nucleolar ribonucleoproteins. RNA 5:1597–1604

Massenet S, Motorin Y, Lafontaine D, Hurt E, Grosjean H, Branlant C (1999) Pseudouridine mapping in the *Saccharomyces cerevisiae* spliceosomal U small nuclear RNAs (snRNAs) reveals that pseudouridine synthase pus1p exhibits a dual substrate specificity for U2 snRNA and tRNA. Mol Cell Biol 19:2142–2154

Massenet S, Pellizzoni L, Paushkin S, Mattaj I, Dreyfuss G (2002) The SMN complex is associated with snRNPs throughout their cytoplasmic assembly pathway. Mol Cell Biol 22:6533–6541

Mattaj I (1986) Cap trimethylation of U snRNA is cytoplasmic and dependent on U snRNP protein binding. Cell 46:905–911

Maxwell ES, Fournier MJ (1995) The nucleolar small RNAs. Annu Rev Biochem 35:897–934

Mayes A, Verdone L, Legrain P, Beggs J (1999) Characterization of Sm-like proteins in yeast and their association with U6 snRNA. EMBO J 18:4321–4331

Meister G, Eggert C, Buhler D, Brahms H, Kambach C, Fischer U (2001) Methylation of Sm proteins by a complex containing PRMT5 and the putative U snRNP assembly factor pICln. Curr Biol 11:1990–1994

Meister G, Eggert C, Fischer U (2002) SMN-mediated assembly of RNPs: a complex story. Trends Cell Biol 12:472–478

Mitchell J, Cheng J, Collins K (1999) A box H/ACA small nucleolar RNA-like domain at the human telomerase RNA 3′-end. Mol Cell Biol 19:567–576

Mouaikel J, Verheggen C, Bertrand E, Tazi J, Bordonné R (2002) Hypermethylation of the cap structure of both yeast snRNAs and snoRNAs requires a conserved methyltransferase that is localized to the nucleolus. Mol Cell 9:891–901

Mouaikel J, Narayanan U, Verheggen C, Matera AG, Bertrand E, Tazi J, Bordonné R (2003) Interaction between the small-nuclear-RNA cap hypermethylase and the spinal muscular atrophy protein, SMN. EMBO Rep 4:616–622

Narayanan A, Speckmann W, Terns R, Terns M (1999a) Role of the box C/D motif in localization of small nucleolar RNAs to coiled bodies and nucleoli. Mol Biol Cell 10:2131–2147

Narayanan A, Lukowiak A, Jady B, Dragon F, Kiss T, Terns R, Terns M (1999b) Nucleolar localization signals of box H/ACA small nucleolar RNAs. EMBO J 18:5120–5130

Narayanan U, Ospina J, Frey M, Hebert M, Matera A (2002) SMN, the spinal muscular atrophy protein, forms a pre-import snRNP complex with snurportin1 and importin beta. Hum Mol Genet 11:1785–1795

Ni J, Tien A, Fournier M (1997) Small nucleolar RNAs direct site-specific synthesis of pseudouridine in ribosomal RNA. Cell 89:565–573

Niewmierzycka A, Clarke S (1999) S-Adenosylmethionine-dependent methylation in *Saccharomyces cerevisiae*. Identification of a novel protein arginine methyltransferase. J Biol Chem 274:814–824

Ohno M, Segref A, Bachi A, Wilm M, Mattaj I (2000) PHAX, a mediator of U snRNA nuclear export whose activity is regulated by phosphorylation. Cell 101:187–198

Omer A, Lowe T, Russell A, Ebhardt H, Eddy S, Dennis P (2000) Homologs of small nucleolar RNAs in Archaea. Science 288:517–522

Ooi S, Samarsky D, Fournier M, Boeke J (1998) Intronic snoRNA biosynthesis in *Saccharomyces cerevisiae* depends on the lariat-debranching enzyme: intron length effects and activity of a precursor snoRNA. RNA 4:1096–1110

Pannone B, Xue D, Wolin S (1998) A role for the yeast La protein in U6 snRNP assembly: evidence that the La protein is a molecular chaperone for RNA polymerase III transcripts. EMBO J 17:7442–7453

Paushkin S, Gubitz A, Massenet S, Dreyfuss G (2002) The SMN complex, an assemblyosome of ribonucleoproteins. Curr Opin Cell Biol 14:305–312

Pelczar P, Filipowicz W (1998) The host gene for intronic U17 small nucleolar RNAs in mammals has no protein-coding potential and is a member of the 5′-terminal oligopyrimidine gene family. Mol Cell Biol 18:4509–4518

Pellizzoni L, Kataoka N, Charroux B, Dreyfuss G (1998) A novel function for SMN, the spinal muscular atrophy disease gene product, in pre-mRNA splicing. Cell 95:615–624

Pellizzoni L, Baccon J, Charroux B, Dreyfuss G (2001) The survival of motor neurons (SMN) protein interacts with the snoRNP proteins fibrillarin and GAR1. Curr Biol 11:1079–1088

Petfalski E, Dandekar T, Henry Y, Tollervey D (1998) Processing of the precursors to small nucleolar RNAs and rRNAs requires common components. Mol Cell Biol 18:1181–1189

Pillai R, Will C, Lührmann R, Schumperli D, Muller B (2001) Purified U7 snRNPs lack the Sm proteins D1 and D2 but contain Lsm10, a new 14 kDa Sm D1-like protein. EMBO J 20:5470–5479

Plessel G, Fischer U, Lührmann R (1994) m₃G cap hypermethylation of U1 small nuclear ribonucleoprotein (snRNP) in vitro: evidence that the U1 small nuclear RNA-(guanosine-N2)-methyltransferase is a non-snRNP cytoplasmic protein that requires a binding site on the Sm core domain. Mol Cell Biol 14:4160–4172

Raker V, Plessel G, Lührmann R (1996) The snRNP core assembly pathway: identification of stable core protein heteromeric complexes and an snRNP subcore particle in vitro. EMBO J 15:2256–2269

Reddy R, Busch H (1988) Small nuclear RNAs: RNA sequences, structure, and modifications. In: Birnstiel ML (ed) Structure and function of major and minor nuclear ribonucleoprotein particles. Springer, Berlin Heidelberg New York, pp 1–37

Ryan D, Stevens S, Abelson J (2002) The 5′ and 3′ domains of yeast U6 snRNA: Lsm proteins facilitate binding of Prp24 protein to the U6 telestem region. RNA 8:1011–1033

Salgado-Garrido J, Bragado-Nilsson E, Kandels-Lewis S, Séraphin B (1999) Sm and Sm-like proteins assemble in two related complexes of deep evolutionary origin. EMBO J 18:3451–3462

Samarsky D, Fournier M (1999) A comprehensive database for the small nucleolar RNAs from *Saccharomyces cerevisiae*. Nucleic Acids Res 27:161–164

Samarsky D, Fournier M, Singer R, Bertrand E (1998) The snoRNA box C/D motif directs nucleolar targeting and also couples snoRNA synthesis and localization. EMBO J 17:3747–3757

Schimmang T, Tollervey D, Kern H, Frank R, Hurt E (1989) A yeast nucleolar protein related to mammalian fibrillarin is associated with small nucleolar RNA and is essential for viability. EMBO J 8:4015–4024

Segault V, Will C, Sproat B, Lührmann R (1995) In vitro reconstitution of mammalian U2 and U5 snRNPs active in splicing: Sm proteins are functionally interchangeable and are essential for the formation of functional U2 and U5 snRNPs. EMBO J 14:4010–4021

Seipelt R, Zheng B, Asuru A, Rymond B (1999) U1 snRNA is cleaved by RNase III and processed through an Sm site-dependent pathway. Nucleic Acids Res 27:587–595

Seraphin B (1995) Sm-like proteins belong to a large family: identification of proteins of the U6 as well as the U1, U2, U4 and U5 snRNPs. EMBO J 14:2089–2098

Seto A, Zaug A, Sobel S, Wolin S, Cech T (1999) *Saccharomyces cerevisiae* telomerase is an Sm small nuclear ribonucleoprotein particle. Nature 40:177–180

Sleeman J, Lamond A (1999) Newly assembled snRNPs associate with coiled bodies before speckles, suggesting a nuclear snRNP maturation pathway. Curr Biol 9:1065–1074

Smith C, Steitz J (1998) Classification of gas5 as a multi-small-nucleolar-RNA (snoRNA) host gene and a member of the 5′-terminal oligopyrimidine gene family reveals common features of snoRNA host genes. Mol Cell Biol 18:6897–6909

Stade K, Ford C, Guthrie C, Weis K (1997) Exportin 1 (Crm1p) is an essential nuclear export factor. Cell 90:1041–1050

Steinmetz E, Conrad N, Brow D, Corden J (2001) RNA-binding protein Nrd1 directs poly(A)-independent 3′-end formation of RNA polymerase II transcripts. Nature 413:327–331

Terns M, Grimm C, Lund E, Dahlberg J (1995) A common maturation pathway for small nucleolar RNAs. EMBO J 14:4860–4871

Tollervey D, Lehtonen H, Jansen R, Kern H, Hurt E (1993) Temperature-sensitive mutations demonstrate roles for yeast fibrillarin in pre-rRNA processing, pre-rRNA methylation, and ribosome assembly. Cell 72:443–457

Toro I, Thore S, Mayer C, Basquin J, Séraphin B, Suck D (2001) RNA binding in an Sm core domain: X-ray structure and functional analysis of an archaeal Sm protein complex. EMBO J 20:2293–2303

Toro I, Basquin J, Teo-Dreher H, Suck D (2002) Archaeal Sm proteins form heptameric and hexameric complexes: crystal structures of the Sm1 and Sm2 proteins from the hyperthermophile Archaeoglobus fulgidus. J Mol Biol 320:129–142

Tyc K, Steitz J (1989) U3, U8 and U13 comprise a new class of mammalian snRNPs localized in the cell nucleolus. EMBO J 8:3113–3119

Tycowski K, Shu M, Steitz J (1996) A mammalian gene with introns instead of exons generating stable RNA products. Nature 379:464–466

Tycowski K, You Z, Graham P, Steitz J (1998) Modification of U6 spliceosomal RNA is guided by other small RNAs. Mol Cell 2:629–638

Urlaub H, Raker V, Kostka S, Lührmann R (2001) Sm protein-Sm site RNA interactions within the inner ring of the spliceosomal snRNP core structure. EMBO J 20:187–196

Valadkhan S, Manley J (2001) Splicing-related catalysis by protein-free snRNAs. Nature 413:701–707

Verheggen C, Mouaikel J, Thiry M, Blanchard J, Tollervey D, Bordonne R, Lafontaine D, Bertrand E (2001) Box C/D small nucleolar RNA trafficking involves small nucleolar RNP proteins, nucleolar factors and a novel nuclear domain. EMBO J 20:5480–5490

Verheggen C, Lafontaine D, Samarsky D, Mouaikel J, Blanchard J, Bordonné R, Bertrand E (2002) Mammalian and yeast U3 snoRNPs are matured in specific and related nuclear compartments. EMBO J 21:2736–2745

Vidovic I, Nottrott S, Hartmuth K, Lührmann R, Ficner R (2000) Crystal structure of the spliceosomal 15.5kD protein bound to a U4 snRNA fragment. Mol Cell 6:1331–1342

Villa T, Ceradini F, Presutti C, Bozzoni I (1998) Processing of the intron-encoded U18 small nucleolar RNA in the yeast Saccharomyces cerevisiae relies on both exo- and endonucleolytic activities. Mol Cell Biol 18:3376–3383

Watkins N, Gottschalk A, Neubauer G, Kastner B, Fabrizio P, Mann M, Lührmann R (1998) cbf5p, a potential pseudouridine synthase and Nhp2p, a putative RNA-binding protein, are present together with Gar1p in all H BOX/ACA-motif snoRNPs and constitute a common bipartite structure. RNA 4:1549–1568

Watkins N, Segault V, Charpentier B, Nottrott S, Fabrizio P, Bachi A, Wilm M, Rosbash M, Branlant C, Lührmann R (2000) A common core RNP structure shared between the small nucleolar box C/D RNPs and the spliceosomal U4 snRNP. Cell 103:457–466

Watkins N, Dickmanns A, Lührmann R (2002) Conserved stem II of the box C/D motif is essential for nucleolar localization and is required, along with the 15.5 K protein, for the hierarchical assembly of the box C/D snoRNP. Mol Cell Biol 22:8342–8352

Will C, Lührmann R (2001) Spliceosomal UsnRNP biogenesis, structure and function. Curr Opin Cell Biol 13:290–301

Wong J, Kusdra L, Collins K (2002) Subnuclear shuttling of human telomerase induced by transformation and DNA damage. Nat Cell Biol 4:731–736

Wu P, Brockenbrough J, Metcalfe A, Chen S, Aris J (1998) Nop5p is a small nucleolar ribonucleoprotein component required for pre-18S rRNA processing in yeast. J Biol Chem 273:16453–16463

Xue D, Rubinson D, Pannone B, Yoo C, Wolin S (2000) U snRNP assembly in yeast involves the La protein. EMBO J 19:1650–1660

Yu Y, Shu M, Steitz J (1998) Modifications of U2 snRNA are required for snRNP assembly and pre-mRNA splicing. EMBO J 17:5783–5795

Zebarjadian Y, King T, Fournier M, Clarke L, Carbon J (1999) Point mutations in yeast cbf5 can abolish in vivo pseudouridylation of rRNA. Mol Cell Biol 19:7461–7472

Intranuclear Pre-mRNA Trafficking in an Insect Model System

Eva Kiesler[1] and Neus Visa[1]

1
The Balbiani Ring Genes of *Chironomus tentans*: a Model System for Studying Gene Expression in Situ

The expression of protein-coding genes in eukaryotic cells is a multi-step process that starts with the synthesis of a pre-messenger RNA (pre-mRNA) in the cell nucleus and finishes with the synthesis of a protein in the cytoplasm. In the nucleus, the pre-mRNA must undergo several processing reactions to become a mature mRNA that is subsequently exported to the cytoplasm. Many different enzymatic activities participate in the gene expression pathway, and the molecules or multi-molecular complexes responsible for these activities have been identified and characterized in considerable detail at the biochemical level (e.g., Ban et al. 2000; Shatkin and Manley 2000; Cramer et al. 2001; Will and Lührmann 2001a). Indeed, our knowledge about the gene expression processes is such that many of the steps of the gene expression pathway can be reproduced individually in vitro. However, nucleocytoplasmic transport can only be envisaged in relation to the cellular architecture, and for this reason the understanding of the transport mechanisms relies mostly on in situ studies. The larval salivary glands of the dipteran *Chironomus tentans* constitute a valuable model system for in situ studies of gene expression, as they offer the possibility of directly visualizing the synthesis, processing and export of a specific transcript: the Balbiani ring (BR) pre-mRNA. The BR genes and their expression in the salivary gland cells have been reviewed in detail previously (e.g., Mehlin and Daneholt 1993; Wieslander 1994; Daneholt 2001). In this chapter, we will describe the BR system, we will discuss the contribution of the BR system to our understanding of mRNA trafficking, and we will discuss the possible implication of nonchromatin nuclear structures in mRNA biogenesis.

[1]Department of Molecular Biology and Functional Genomics, Stockholm University, 10961 Stockholm, Sweden. Tel. +46 8 164111; Fax. +46 8 166488, e-mail: neus.visa@molbio.su.se

Progress in Molecular and Subcellular Biology
P. Jeanteur (Ed.): RNA Trafficking and Nuclear Structure Dynamics
© Springer-Verlag Berlin Heidelberg 2003

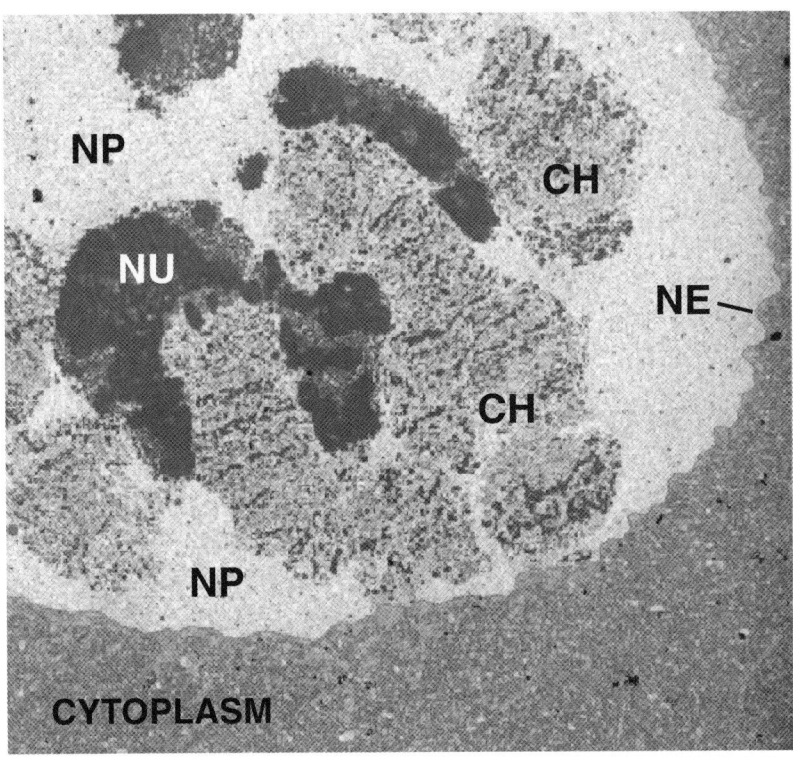

Fig. 1. Electron micrograph showing the nucleus of a salivary gland cell from *C. tentans*. *CH* chromosome, *NE* nuclear envelope, *NP* nucleoplasm, *NU* nucleolus. The *bar* represents 5 μm

1.1
The Active Balbiani Ring Genes

The larval salivary gland cells of *C. tentans* possess a polytene nucleus with four giant chromosomes (reviewed by Hägele 1975; Case and Daneholt 1977). At a gross level, four major structures or compartments can be easily identified in these nuclei: the nuclear envelope, the polytene chromosomes, the nucleoli and the interchromatin or nucleoplasm (Fig. 1). Each polytene chromosome consists of thousands of chromatids arranged in register, and the variable degree of chromatin condensation along each chromosome results in a characteristic banding pattern that allows the identification of specific loci. At transcriptionally active sites, the chromatin is expanded or puffed.

Chromosome IV contains two puffs of exceptional size, referred to as Balbiani rings (BR) 1 and 2, which originate as a consequence of the expression of the BR genes (reviewed by Wieslander 1994). The pre-mRNA synthesized in BR1 and BR2 is unusually long (35–40 kb). Upon transcription, the BR pre-

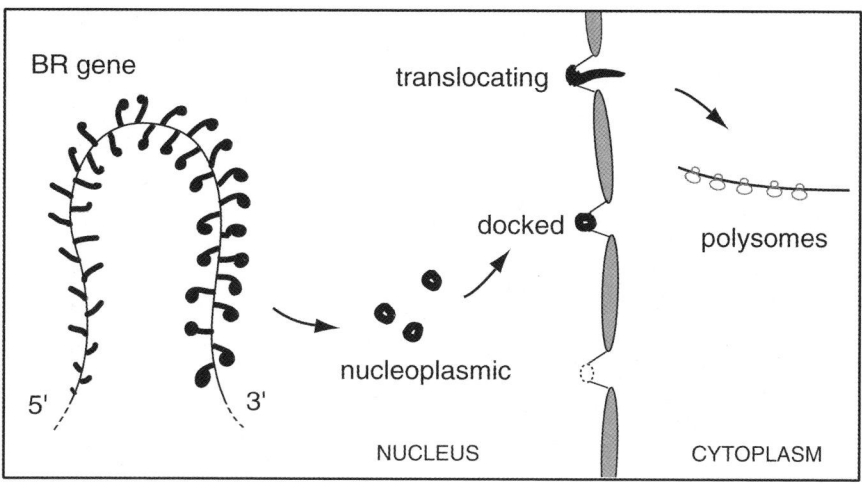

Fig. 2. Schematic representation of BR (pre)mRNP particles showing successive stages of synthesis and assembly of the pre-mRNP at the BR gene, transport through the nucleoplasm, docking at the NPC, translocation to the cytoplasm and translation in polysomes. The polarity of the gene (5′ to 3′) is indicated

mRNA is associated with proteins and assembled into a large ribonucleoprotein (RNP) complex called the BR RNP particle. Due to their structure and extraordinary dimensions, the BR RNP particles can be unambiguously identified in the transmission electron microscope (TEM), and their assembly, transport and disassembly can be directly studied in situ (reviewed by Mehlin and Daneholt 1993; Daneholt 2001). Nascent BR pre-mRNA molecules are rapidly packaged into growing RNP fibers that can be observed along the BR genes (Fig. 2). As transcription progresses, the growing RNP fibers gradually increase in length, which confers a distinct polarity to the active BR transcription unit (Skoglund et al. 1983). When transcription is completed, the BR particles are released from the chromosome and can be seen in the nucleoplasm as RNP granules with a diameter of about 50 nm. The BR RNP granules move towards the nuclear envelope, where they unfold and become elongated during translocation through the nuclear pores (Mehlin et al. 1992).

1.1.1
Processing of BR pre-mRNA

In spite of their unusual length, the BR pre-mRNAs have all the features of a typical pre-mRNA and undergo normal pre-mRNA processing. The 5′ end is capped, and the nuclear cap binding complex can be detected in situ on the nascent pre-mRNP early after transcription initiation (Visa et al. 1996a). The

BR pre-mRNA is cleaved and polyadenylated at the 3' end (Baurén et al. 1998), and it contains four short introns that are spliced prior to export of the mRNA to the cytoplasm (Baurén and Wieslander 1994). Kinetic experiments in vivo have shown that the BR mRNA is efficiently exported to the cytoplasm and is highly stable, with an average half-life of about 20 h, as expected for a gene that is actively expressed (Edström et al. 1978 and references therein).

Studies in the last decade have shown that the different steps of the gene expression pathway are not independent of each other, but are tightly co-ordinated in the living cell by a complex network of molecular interactions (reviewed by Maniatis and Reed 2002). Studies carried out in the BR system have provided clear evidence for a coupling between transcription and splicing in vivo (Baurén and Wieslander 1994; Wetterberg et al. 2001), and between transcription termination, splicing and 3'-end processing (Baurén et al. 1998). In these studies, BR (pre-)mRNP particles have been isolated from different nuclear compartments by microdissection, and the splicing status of the RNA in the particles has been analyzed by RT-PCR. This experimental approach has revealed that the three short introns located close to the 5' end of BR1 pre-mRNA are spliced mostly cotranscriptionally (Baurén and Wieslander 1994). The fourth intron of BR1 pre-mRNA, which is located near the 3' end, is removed also at the transcription site, but after 3'-end cleavage in about 80% of the BR pre-mRNA molecules; in the remaining 20% of the transcripts, intron four is removed after the transcript has left the gene (Baurén et al. 1998).

1.2
Proteins Associated with the BR (pre-)mRNA

One of the major contributions of the BR system to the study of gene expression in situ is the possibility of analyzing the association of defined RNA-binding proteins with the BR RNP particles during successive stages of synthesis, maturation, and export using immuno-electron microscopy (IEM; reviewed by Daneholt 2001). Thus, it is possible in the BR system to determine at what step of the gene expression pathway a specific protein interacts with the (pre-)mRNA. The interest of this IEM approach lies in the assumption that the proteins involved in the gene expression pathway are remarkably con-served through evolution, so that the results obtained in *C. tentans* can be extrapolated to other organisms, ideally to all metazoans. For this reason, considerable effort has been made to identify nuclear proteins of *C. tentans*, verify their homology to proteins in other species, and carry out IEM assays in the BR system. Several proteins have been analyzed in this way (Table 1). Full-length cDNA clones have been isolated and sequenced for most of these proteins, and orthologs in other species have been assigned on the basis of cDNA sequence comparisons. Only in a few cases (PABPN1, CBP20 and REF) has the identification of the *C. tentans* proteins been based on the use of specific antibodies raised against homologous proteins from other species.

Table 1. BR RNA-associated proteins

Protein[a]	Human orthologs(s)	Putative function(s)[b]	Association with BR RNP[c]	References[d]
hrp23/RSF1	??	Packaging/splicing	From gene to NPC	Sun et al. (1998)
hrp36	hnRNP A/B	Packaging	From gene to polysomes	Visa et al. (1996b)
hrp45	ASF/SF2	Packaging/splicing	From gene to NPC	Alzhanova-Ericsson et al. (1996)
hrp84[e]	DBX	Translation initiation	From gene to polysomes	
RAE1	mrnp40	Nuclear export	NPC proximity	Sabri et al. (2001)
Dbp5	Dbp5	Nuclear export	From gene to NPC	Zhao et al. (2002)
HEL	UAP56	Splicing/export	From gene to NPC	Kiesler et al. (2002)
REF	Aly/REF	Nuclear export	From gene to NPC	Kiesler et al. (2002)
PABPN1	PABP1	Poly(A)-binding	From gene to NPC	Bear et al. (2003)
CBP20	CBP20	Cap-binding	From gene to NPC	Visa et al. (1996a)
hrp65	PSP1, PSF, p54[nrb]	Splicing/retention/transcription		Miralles et al. (2000)
p2D10	TFIIICα	Transcription/possible post-Transcriptional		Sabri et al. (2002)
actin	actin	Transcription/possible post-Transcriptional	From gene to polysomes	Percipalle et al. (2001)

[a]Name of the protein in *C. tentans*
[b]Based on known function of orthologs in other species and on localization of the protein in *C. tentans*
[c]As shown by IEM
[d]Reference to the *C. tentans* protein
[e]D. Nashchekin, J. Zhao, N. Fomproix, B. Ivarsson, N. Visa and B. Daneholt (unpubl. data)

As shown in Table 1, eleven proteins have been identified that associate with the BR RNP particles, and it is possible to identify a mammalian ortholog for most of them. Some of the proteins found in the BR RNP particles, such as hrp23, hrp36 and hrp45, are abundant proteins present in multiple copies along the BR pre-mRNP, and it is likely that these proteins play a role in the packaging of the BR pre-mRNA. In some cases, the association is restricted to the nucleus, as in the case of hrp23 (Sun et al. 1998) and hrp45 (Alzhanova-Ericsson et al. 1996), whereas other proteins such as hrp36 are exported with the BR mRNA to the cytoplasm and remain associated with the BR mRNA during translation (Visa et al. 1996b).

Not all the proteins located at the BR RNP particle are packaging proteins. Some of them, such as CBP20 and PABPN1, are bound to specific structural elements of the BR (pre-)mRNA and are likely to play specialized functions in mRNA biogenesis.

It is also interesting to note that, although the proteins found in the BR RNP particle are of different types and are likely to play specific functions in the gene expression pathway, most of them become associated with the BR pre-mRNA as early as during transcription. Based on this observation, it has been proposed that the fate of the mRNA is determined concomitantly with transcription, during the early assembly of the RNP at the gene (Daneholt 2001).

1.2.1
DExD/H Box Proteins in the BR RNP Particle

Three RNA helicases of the DExD/H box family (Tanner and Linder 2001) have been found associated with the BR RNP particle: hrp84, Dbp5 and HEL (Table 1). Hrp84 is a putative RNA helicase that belongs to the PL10 subfamily of DExD/H box helicases (J. Zhao, D. Nashchekin, N. Visa and B. Daneholt, unpubl. data), which includes several proteins thought to be involved in translation initiation: the human DBX, the mouse PL10 and the yeast Ded1p (Leroy et al. 1989; Gururajan et al. 1991; Mamiya and Worman 1999). IEM experiments have shown that hrp84 is added co-transcriptionally to the nascent BR pre-mRNA, and that it remains on the BR RNP particle during export across the nuclear membrane and all the way into polysomes, where it is thought to play a role in translation initiation.

Dbp5 is an essential mRNA export factor in yeast and in vertebrates (Snay-Hodge et al. 1998; Tseng et al. 1998; Schmitt et al. 1999). Due to its predominantly cytoplasmic distribution and to its location at the cytoplasmic fibrils of the nuclear pore complex (NPC), Dbp5 is thought to function during late steps of mRNA export (Schmitt et al. 1999; Strahm et al. 1999). The *C. tentans* ortholog of Dbp5 has been recently characterized (Zhao et al. 2002). IEM of Dbp5 in *C. tentans* salivary gland cells has confirmed the abundance of Dbp5 in the cytoplasm, and revealed that a fraction of Dbp5 is present at the nucleoplasmic side of the NPC. As in the case of hrp84, Dbp5 is incorporated into the growing BR RNP fibril early during transcription, and remains associated with the BR RNP particle during transport to the cytoplasm (Zhao et al. 2002). These observations support the idea that Dbp5 is involved in the remodeling of the mRNPs on their arrival at the cytoplasm, and that Dbp5 may play a role in coupling mRNP export with translation (e.g., Tseng et al. 1998; Schmitt et al. 1999).

The third putative helicase of the DExD/H box family present in the BR RNP article is HEL, the dipteran ortholog of mammalian UAP56 and yeast Sub2p (Kiesler et al. 2002 and references therein). UAP56 was initially identified as an essential splicing factor (Fleckner et al. 1997), and recent studies have shown that this protein, as well as its orthologs in yeast and insects, is also essential for the export of mRNA (reviewed by Linder and Stutz 2001). HEL/

UAP56 associates with the export factor Aly/REF/Yra1 (reviewed by Conti and Izaurralde 2001), a component of a stable protein complex deposited onto the mRNA upon splicing: the exon-junction complex (EJC; reviewed by Kim and Dreyfus 2001). The role of HEL/UAP56 in splicing and its binding to Aly/REF/ Yra1 have led to the proposal that HEL binds to the pre-mRNA during splicing and recruits the EJC to the spliced mRNA (Luo et al. 2001; Stässer and Hurt 2001). However, HEL and Sub2p are also required for the export of intron-less pre-mRNAs (Gatfield et al. 2001; Jensen et al. 2001a; Strässer and Hurt 2001), which indicates that the role of HEL in pre-mRNA export is not necessarily linked to splicing. This suggestion is supported by investigations of HEL in *C. tentans*. Quantitative IEM studies have shown that HEL binds progressively to the BR pre-mRNA as transcription proceeds, regardless of the position of introns along the pre-mRNA (Kiesler et al. 2002). Subsequently, HEL remains associated with the BR particle during transport through the nucleoplasm, and is released from the BR particle during translocation through the NPC (Kiesler et al. 2002).

It has been proposed that the DExH/D proteins regulate the organization of the RNP complexes by altering either RNA – RNA or RNA – protein interactions, and that these proteins could thus act as RNP remodelling factors at different stages of mRNA biogenesis (Will and Lührmann 2001b). In the case of the BR particles, hrp84, Dbp5 and HEL are co-transcriptionally assembled with the pre-mRNP, but are released at different time-points during or after transport of the BR RNP particle to the cytoplasm. More research is needed to determine the time-points at which each helicase is required, the specific substrates of these helicases, and the events that trigger their ATP-dependent activities.

2
The Movement of the (pre)mRNP Particles from the Gene to the NPC

Several factors that mediate export of mRNA from the nucleus to the cytoplasm have been discovered in the last few years, mainly by studies in yeast, *Drosophila* and mammals. The pathways by which such factors mediate export of mRNPs to the cytoplasm have also begun to be elucidated (reviewed by Lei and Silver 2002; Reed and Hurt 2002). However, there are crucial mechanistic aspects concerning the intranuclear trafficking of pre-mRNPs that are still poorly understood. For instance, it is not known how or when the pre-mRNP becomes destined for export, or whether the transported mRNPs are freely diffusable throughout their journey through the nucleoplasm, or whether they interact with nuclear structures.

The BR model system of *C. tentans* provides a unique opportunity to investigate directly whether (pre)mRNP particles interact with nuclear substructures after being released from the gene. The three-dimensional (3D) structure of the BR RNP particle has been determined both in situ (Skoglund et al. 1986) and after separation in sucrose gradients (Lönnroth et al. 1992) using electron tomography (ET), which enables the identification of BR particles in TEM preparations based on structural features. In addition, the organization of chromatin into polytene chromosomes in the salivary gland cell nuclei certifies that any structures observed within the large interchromatin space are non-chromatin structures.

2.1
Three-Dimensional-Analysis of Nucleoplasmic BR RNP Particles by Electron Tomography

We decided to study the intranuclear transport of mRNPs from the gene to the NPC by directly visualizing BR RNP particles in transit from the gene to the nuclear envelope and determining whether these nucleoplasmic particles showed any morphological sign of physical binding to other structures. Early TEM observations had suggested that BR particles in transit from the gene to the NPC are associated with thin fibers that extend into the surrounding nucleoplasm (Fig. 3A). To confirm these initial observations and to obtain reliable 3D information about possible interactions between the BR RNP particles and other nuclear structures, the nucleoplasmic BR RNP particles were analyzed using cryosections of fixed salivary gland cells and ET. Using this method, Miralles and co-workers (2000) showed that approximately one third of the nucleoplasmic BR particles do not show any sign of binding interactions; the morphology of these so-called free BR particles was thus compatible with a free diffusion model. However, an equally large fraction of nucleoplasmic BR RNP particles was found in direct contact with thin fibers referred to as connecting fibers, or CFs (Fig. 3B). The CFs have a diameter of approximately 7 nm and a length in the 15–50 nm range. In some cases the CFs merged into larger structures of unknown function called fibrogranular clusters, or FGCs. Although the function of these nucleoplasmic structures remains to be elucidated, it is clear that the binding of BR RNP particles to large nonchromatin structures such as CFs and FGCs imposes restrictions on the movement of the BR particles. Thus, the BR RNP particles attached to CFs and FGCs cannot diffuse freely in the nucleoplasm, at least not at rates expected for free RNPs (Miralles et al. 2000).

The ET studies of Miralles et al. (2000) were designed to analyze short fibers connected to the BR RNP particles and did not provide information about large nucleoplasmic structures such as FGCs. Studies are now in progress to determine the size and structure of the FGCs, and to establish whether each FGC is an isolated structure or part of a large fibrous meshwork extending throughout the nucleoplasm (N. Visa and U. Skoglund, unpubl. data).

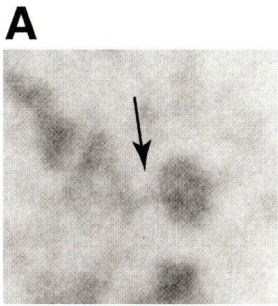

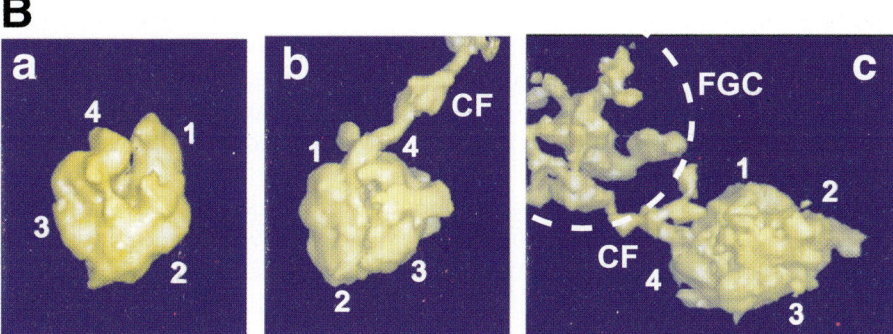

Fig. 3. 3D-reconstructions of nucleoplasmic BR particles. **A** Electron micrograph showing a nucleoplasmic BR particle in a thin cryosection of a salivary gland observed in the TEM. The BR particle appears as a dense granule with a diameter of about 50 nm. The *arrow* points at a fiber that contacts the BR particle and extends into the surrounding nucleoplasm. The *bar* represents 100 nm. **B** Three examples of nucleoplasmic BR RNP particles analyzed by electron tomography. The numbers on the images *1* to *4* refer to the domains of the BR particles (Skoglund et al. 1986). The BR particle shown in **a** is apparently free, not in contact with any other structure. The BR particles in **b** and **c** are in contact with connecting fibers *CF*s that extend into the nucleoplasm. The fiber in **c** is continuous with a fibrogranular cluster *FGC* located in the upper left corner of the image. The *bar* represents 25 nm

2.2
Protein Components of CFs and FGCs

The CFs could be labeled by IEM with a monoclonal antibody raised against nucleoplasmic proteins from *C. tentans* salivary gland cells. This antibody enabled the identification of a protein component of the CFs: the hrp65 protein (Miralles et al. 2000). Hrp65 is a basic protein of 65 kDa that contains two classical RNA-binding domains (RBDs). Recombinant hrp65 can form oligomers in vitro, which has led to the proposal that hrp65 is involved in the assembly of the CFs in vivo (Kiesler et al. 2003).

Hrp65 is evolutionarily conserved in metazoans, and the mammalian orthologs of hrp65 have been implicated in different steps of gene expression (see next section).

It is not clear whether the CFs contain RNA or only protein, but it has been shown by IEM that the CFs have a specific protein composition different from that of the BR RNP particles: CFs are labeled by an anti-hrp65 antibody, but not by antibodies against the abundant proteins hrp23, hrp45 and hrp36 (Table 1 and Miralles et al. 2000). The CFs are not labeled by antibodies against the cap-binding proteins or against the polyA-binding protein (Visa et al. 1996a; Bear et al. 2003), which rules out the possibility that CFs correspond to a terminal portion of the BR RNP fiber in an extended conformation.

The only protein component of the FGCs identified so far is the *C. tentans* protein p2D10 (Table 1), which was initially identified as a nucleoplasmic protein and found to share structural similarity with known TFIIIC-α proteins from other organisms (Sabri et al. 2002). Although the location of p2D10 to FGCs has not been demonstrated by immuno-ET, conventional IEM results strongly suggest that this protein is preferentially located on FGCs (Sabri et al. 2002).

The identification of additional components of CFs and FGCs is currently in progress and we hope that this will help us to understand the function of these structures and the molecular basis for their interaction with the BR RNP particles.

2.2.1
Hrp65 and Other DBHS Proteins

Sequence analysis reveals that hrp65 is highly similar to a group of RNA-binding proteins characterized by the presence of a DBHS domain (Dong et al. 1993). This group of proteins includes mammalian PSP-1, PSF, p54nrb/NonO and *Drosophila* NonA/Bj6 (Jones and Rubin 1990; Dong et al. 1993; Patton et al. 1993; Fox et al. 2002). The two most studied mammalian DBHS proteins, PSF and p54[nrb], are involved in a variety of nuclear processes (reviewed by Shav-Tal and Zipori 2002). PSF functions in pre-mRNA splicing (e.g., Patton et al. 1993; Peng et al. 2002) and mediates transcriptional regulation together with p54[nrb] (Basu et al. 1997; Yang et al. 1997; Mathur et al. 2001; Sewer and Waterman 2002; Sewer et al. 2002; Emili et al. 2002). The PSF – p54[nrb] complex also mediates nuclear retention of inosine-rich viral RNAs (Zhang and Car-michael 2001). It is difficult to infer the function of hrp65 from the function of the mammalian DHBS proteins, although the role of the PSF – p54[nrb] complex in nuclear retention of viral RNA may be related to the function of hrp65 in CFs.

The DBHS domain is composed of an upstream part with two RBDs, and a conserved downstream part able to mediate oligomerization of hrp65 and the heterodimerization between DBHS proteins (Peng et al. 2002; Kiesler et al. 2003).

2.2.2
The hrp65 Isoforms: Specialized Roles in Gene Expression?

Three isoforms of hrp65 that originate by alternative splicing from a single pre-mRNA have been characterized (Miralles and Visa 2001). Two of the hrp65 isoforms, hrp65-1 and hrp65-2, are ubiquitously expressed during larval and adult development of *C. tentans*, while the hrp65-3 isoform has only been detected in a cell line of embryonic origin (Miralles and Visa 2001).

Two different antibodies against hrp65 failed to label the active BR genes, showing that hrp65 does not interact with the BR RNP particles co-transcriptionally (Miralles et al. 2000). However, the isoform hrp65-2 has recently been detected at the BR genes using an isoform-specific antibody. Furthermore, hrp65-2 plays a role in transcription regulation (Percipalle et al. 2003). Further work is required to establish the function(s) of each hrp65 isoform, but the available results so far suggest that hrp65-2 is a less abundant isoform involved in transcription regulation, whereas hrp65-1 appears to be the predominating nuclear isoform located in CFs.

2.3
Intranuclear Movement of (pre-)mRNPs

The most interesting question that emerges from the ET analysis of nucleoplasmic BR RNP particles reported above is the functional significance of the observed interactions between BR RNP particles and nonchromatin nucleoplasmic structures. According to early kinetic studies (Edström et al. 1978), between 80 and 100% of the synthesized BR RNP particles are exported to the cytoplasm, and thus the BR particles that interact with CFs and FGCs must also be exported. This implies that the different types of BR RNP particles detected in the ET studies of Miralles et al. (2000) should be regarded as different intermediates in the processing and export pathway.

A model that summarizes the results of the ET studies is presented in Fig. 4. The existence of BR RNP particles bound to CFs and FGCs indicates that BR particles are not always free to diffuse in the nucleoplasm, but interact transiently with CFs and FGCs. There are at least two possible ways to resolve the association of BR particles with FGCs. The BR particles may be transiently retained by CFs and FGCs until they are eventually released into the nucleoplasm, where they can diffuse towards the NPC to be exported to the cytoplasm (steps 3 and 4 in Fig. 4). Alternatively, the attached BR particles may never be released back into the nucleoplasm, but brought to the proximity of the NPC

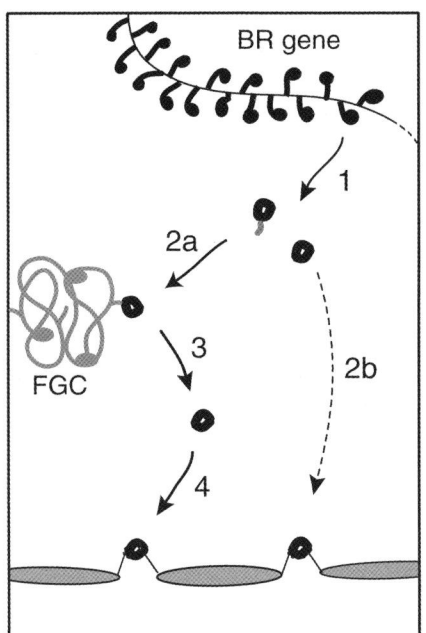

Fig. 4. A model for the intranuclear movement of the BR (pre)mRNP particle. The newly synthesized BR pre-mRNP particle is released into the nucleoplasm *1*. The nucleoplasmic BR particle binds to a fibrogranular cluster or FGC *2a*. The BR particle is subsequently released into the nucleoplasm *3* and can reach the NPC by free diffusion *4*. Some BR particles may never become attached to FGCs and may be able to diffuse directly from the gene to the NPC *2b*

by guided transport along FGCs. In this second case, the binding of BR RNP particles to FGCs would not impair the movement of the mRNP, but actively promote its transport in a directional manner.

Several kinetic studies argue against directional mRNA movement. First, a TEM investigation performed after pulse-labeling of BR particles with bromo-dUTP has shown that the majority of nucleoplasmic BR particles move away from the gene in a random, nondirectional manner (Singh et al. 1999), as opposed to what would be expected if BR particles were actively transported to the NPC by a guided mechanism.

In a second series of experiments, nuclear poly(A)$^+$ RNA was visualized in mammalian cells by hybridization with fluorescent oligo(dT) probes, and the movement of the labeled mRNA in vivo was analyzed by either fluorescence correlation spectroscopy or high-speed imaging microscopy (Politz et al. 1998, 1999). The overall properties of mRNP movement detected in these experiments were compatible with a diffusion model (reviewed by Politz and Pederson 2000), but two main populations of poly(A)$^+$ RNA were detected that differed in their diffusion rates: a major population showed diffusion rates comparable to those of an average-sized mRNP in solution, whereas another population, corresponding to approximately one third of the total poly(A)$^+$ RNA, moved at much lower rates (Politz et al. 1998). Politz and co-workers proposed that mRNPs move in the nucleoplasm by diffusion, but that there may be a fraction of transcripts tethered to large macromolecular structures

(Politz et al. 1998). Although the nature of this slowly moving poly(A)$^+$ RNA in mammalian cells is unknown, it is tempting to speculate that it corresponds to (pre)mRNPs transiently retained by nucleoplasmic structures prior to export to the cytoplasm. The existence of two mRNP populations with different diffusion coefficients in mammalian cells agrees well with the finding of different types of BR RNP particles in C. tentans, and supports the view that binding of BR particles to CFs and FGCs slows the movement of the BR particles.

Finally, although the function of hrp65 is not understood, what is known about hrp65 and its orthologs in mammalian cells is more consistent with a role in pre-mRNA processing and/or nuclear RNA retention than one in active transport (see above). For these reasons, we favor the view that the BR particles associated with CFs and FGCs are retained in the nucleus, or move at a considerably lower speed than free nucleoplasmic BR particles.

Another relevant aspect that has been discussed in relation to the intranuclear mobility of ribonucleoprotein particles is the energy requirement (recently reviewed by Carmo-Fonseca et al. 2002). In a recent study, Carmo-Fonseca and co-workers have fused mRNA-binding proteins to GFP, and used the GFP-fusions as reporters for in vivo measurements of mRNP movement by fluorescence recovery after bleaching (FRAP). These experiments have revealed that energy depletion reduces the mobility of RNA-bound reporter proteins (Calapez et al. 2002).

A possible interpretation that reconciles the data from the BR system, the energy requirement shown by Calapez et al. (2002), and the diffusion models derived from the kinetic studies reported above, is that the movement of (pre)mRNP particles from the gene to the NPC is driven by diffusion, but is slowed by transient interactions between the RNP and nonchromatin nucleoplasmic structures (Fig. 4). Releasing this retention would be the energy-dependent step in the process, not the movement itself. Daneholt and co-workers have estimated that the minimum diffusion coefficient of intranuclear BR particles is approximately 0.1 $\mu m^2/sec$, a value that is compatible with free diffusion through the nucleoplasm (Singh et al. 1999). However, this diffusion coefficient calculated from experimental data is approximately ten times lower than the predicted diffusion coefficient for a 50-nm particle diffusing freely in solution (Singh et al. 1999). The difference between calculated and predicted coefficients could be due, at least in part, to transient retention events.

2.4
Nuclear Retention of mRNA

What could be the functional significance of a transient nuclear retention? Several studies have led to the proposal that nuclear retention is part of a checkpoint mechanism to assess the quality of the mRNA before export to the cytoplasm. In mammalian cells, inefficiently processed pre-mRNAs or pre-mRNAs containing premature termination codons are retained at or near the site of transcription (Custodio et al. 1999; Muhlemann et al. 2001). A similar phenomenon has been reported in yeast cells, where transcripts that fail to acquire a poly(A) tail are also retained in the nucleus (Jensen et al. 2001b; Hilleren et al. 2001). Quality control and retention mechanisms are thought to function to prevent not only the expression of aberrant genes, but also the premature export of unprocessed mRNAs. The retention mechanisms are not known, but transient binding of the unprocessed pre-mRNP to nondiffusible structures inside the nucleus is probably the easiest way to achieve such a retention.

The studies mentioned above show that the retention of unprocessed or incorrectly processed transcripts occurs at or near the transcription sites, as judged by immunofluorescence microscopy. However, it is not known whether transcription is arrested so that the retained transcripts remain attached to the gene template, or whether the transcripts are released from the gene, but tethered to neighboring nucleoplasmic structures.

In the case of the BR genes, binding of the BR RNP to nucleoplasmic structures is not restricted to the proximity of the gene. However, the differences observed in the distance between the genes and the retention sites may be due to the different organization of the cell types compared. Indeed, the ratio of nuclear volume to DNA content is approximately two times larger in the giant salivary gland cells of *C. tentans* than in a typical mammalian cell, which implies that the nucleoplasmic components are *diluted* in a much larger volume in the *C. tentans* polytene nuclei. Thus, it is conceivable that the FGC counterparts of mammalian cells are spatially closer to the transcription sites. In any case, more research is needed to determine whether the binding of BR RNP particles to CFs and FGCs is related to mRNA quality control mechanisms or to sorting of processed and unprocessed transcripts.

2.5
CFs and FGCs in Mammalian Cells?

The acquisition of polyteny is a very common process in differentiated insect tissues. For instance, most larval and some adult tissues of *Drosophila* possess polytene nuclei (Ashburner 1970). The basic genetic mechanisms in *Drosophila* have been studied in detail at the molecular level, and no indication has been found for the existence of gene expression mechanisms specific to polytene nuclei. Thus, we believe that the basic mechanisms of gene expression are similar in polytene cells to those in diploid cells, and that the relationships between nuclear structure and gene function are intrinsically conserved among eukaryotes. It is thus relevant to ask whether the FGCs of *C. tentans* are conserved structures with counterparts in other organisms.

The IGCs described in a variety of organisms (reviewed by Puvion and Puvion-Dutilleul 1996; Spector 1996; Misteli 2000) may be functionally equiv-

alent to the FGCs of *C. tentans*. IGCs are clusters of 20-nm granules intercon-
nected by thin fibers (Monneron and Bernhard 1969). IGCs contain a high
concentration of splicing factors that are recruited to active genes in a transcription-dependent manner, and it has been proposed that IGCs are storage
sites for splicing components (Misteli et al. 1997). However, experiments from
several laboratories have shown that IGCs also contain poly(A)$^+$ RNA (e.g., Visa
et al. 1993), and transcripts from several protein-coding genes have been
specifically mapped to IGCs by in situ hybridization (e.g., Besse and Puvion-
Dutilleul 1995; Xing et al. 1995; Bridge et al. 1996; Melcak et al. 2000). Based
on this type of observations, Raska and co-workers have proposed that transcripts that have not been co-transcriptionally spliced may move to IGCs to be
spliced post-transcriptionally (Melcak et al. 2000, 2001). In summary, IGCs
share some features with FGCs: both structures show a fibrogranular texture
and both associate transiently with (pre)mRNAs. The composition of IGCs has
been determined by mass spectrometry and peptide microsequencing (Mintz
et al. 1999), allowing the possible identity between IGCs and FGCs to be tested
experimentally.

Trp structures in the nuclear interior also share common features with FGCs
(Zimowska and Paddy 2002 and references therein). Tpr is a coiled-coil protein
that builds long filaments that extend from the NPC into the nucleoplasm
(Cordes et al. 1997). At least in insect cells, Tpr is found not only at the nuclear
periphery, but also in the nuclear interior, and recent studies in *Drosophila*
have shown that the intranuclear pool of Trp forms dynamic structures that
react to changes in the transcriptional activity of the cell (Zimowska and Paddy
2002). Interestingly, overproduction of Mlp1, a yeast ortholog of Tpr, results
in nuclear accumulation of poly(A)$^+$ RNA (Kosova et al. 2000). Furthermore,
Mlp1 and Mlp2 interact genetically with Yra1 and can be co-immunoprecipi-
tated with Yra1p and several other mRNA export factors, including Sub2p and
Mex67p, the yeast orthologs of HEL/UAP56 and TAP/NXF1, respectively (F.
Stutz, unpubl. res.). In summary, these observations suggest that Mlp1 and
Mlp2 interact with poly(A)$^+$ RNA in the cell nucleus and participate in nuclear
retention of pre-mRNA. It remains to be seen whether Tpr has the same role
in metazoans and whether it forms intranuclear structures similar to the FGCs
observed in *C. tentans*.

3
Conclusions

The synthesis and nucleocytoplasmic transport of a specific pre-mRNP particle, the BR RNP particle, can be directly visualized in the salivary gland cells
of the dipteran *Chironomus tentans* using TEM. Thanks to the possibility of
identifying BR particles in situ in a chromatin-free nucleoplasm, the 3D-
structural study of nucleoplasmic BR particles has provided important

information about the intranuclear trafficking of (pre)mRNP particles. The following conclusions can be drawn:

1. BR RNP particles move by diffusion from the gene to the NPC, but interact transiently with nonchromatin nucleoplasmic structures referred to as CFs and FGCs.
2. The composition of CFs is different to that of the BR RNP particles and is characterized by the presence of hrp65 and the absence of abundant hrps such as hrp23, hrp36 and hrp45.
3. A model for the intranuclear trafficking of RNPs has been presented that reconciles the transient binding of BR RNP particles to nucleoplasmic structures, the intranuclear diffusion of (pre)mRNPs, and the energy-dependent movement of RNPs. According to this model, the RNP particles diffuse through the nucleoplasm, become retained by binding to nonchromatin structures, and are finally released in an energy-dependent manner.
4. The transient retention of RNPs may be related to sorting and/or processing of unprocessed transcripts.
5. IGCs and internal Tpr structures share features with FGCs and may be functionally equivalent to FGCs. Further research is needed to characterize all these structures at the molecular level and to understand their involvement in mRNA biogenesis.

Acknowledgments. We thank F. Stutz and B. Daneholt for communicating unpublished observations, and G.W. Farrants for language editing. Our research is supported by grants from the Swedish Research Council, the Åke Wiberg Foundation and the Carl Trygger Foundation.

References

Alzhanova-Ericsson AT, Sun X, Visa N, Kiseleva E, Wurtz T, Daneholt B (1996) A protein of the SR family of splicing factors binds extensively to exonic Balbiani ring pre-mRNA and accompanies the RNA from the gene to the nuclear pore. Genes Dev 10:2881–2893

Ashburner M (1970) Function and structure of polytene chromosomes during insect development. Adv Insect Physiol 7:1–95

Ban N, Nissen P, Hansen J, Moore PB, Steitz TA (2000) The complete atomic structure of the large ribosomal subunit at 2.4 A resolution. Science 289:905–920

Basu A, Dong B, Krainer AR, Howe CC (1997) The intracisternal A-particle proximal enhancer-binding protein activates transcription and is identical to the RNA- and DNA-binding protein p54nrb/NonO. Mol Cell Biol 17:677–686

Baurén G, Wieslander L (1994) Splicing of Balbiani ring 1 gene pre-mRNA occurs simultaneously with transcription. Cell 76:183–192

Baurén G, Belikov S, Wieslander L (1998) Transcriptional termination in the Balbiani ring 1 gene is closely coupled to 3'-end formation and excision of the 3'-terminal intron. Genes Dev 12:2759–2769

Bear DG, Fomproix N, Soop T, Kylberg K, Björkroth B, Masich S, Daneholt B (2003) Nuclear poly(A) binding protein is associated with RNA polymerase II from the start of transcription and accompanies the released transcript to the nuclear pore Exp Cell Res 286:332–344

Besse S, Puvion-Dutilleul F (1995) Anchorage of adenoviral RNAs to clusters of interchromatin granules. Gene Expr 5:79–92

Bridge E, Riedel KU, Johansson BM, Pettersson U (1996) Spliced exons of adenovirus late RNAs colocalize with snRNP in a specific nuclear domain. J Cell Biol 135:303–314

Calapez A, Pereira HM, Calado A, Braga J, Rino J, Carvalho C, Tavanez JP, Wahle E, Rosa AC, Carmo-Fonseca M (2002) The intranuclear mobility of messenger RNA binding proteins is ATP-dependent and temperature sensitive. J Cell Biol 159:795–805

Carmo-Fonseca M, Platani M, Swedlow JR (2002) Macromolecular mobility inside the cell nucleus. Trends Cell Biol 12:491–495

Case ST, Daneholt B (1977) Cellular and molecular aspects of genetic expression in Chironomus salivary gland cells. In: Paul J (ed) International review of biochemistry. Univ Park Press, Baltimore, pp 45–77

Conti E, Izaurralde E (2001) Nucleocytoplasmic transport enters the atomic age. Curr Opin Cell Biol 13:310–319

Cordes VC, Reidenbach S, Rackwitz HR, Franke WW (1997) Identification of protein p270/Tpr as a constitutive component of the nuclear pore complex-attached intranuclear filaments. J Cell Biol 136:515–529

Cramer P, Bushnell DA, Kornberg RD (2001) Structural basis of transcription: RNA polymerase II at 2.8 angstrom resolution. Science 292:1863–1876

Custodio N, Carmo-Fonseca M, Geraghty F, Pereira HS, Grosveld F, Antoniou M (1999) Inefficient processing impairs release of RNA from the site of transcription. EMBO J 18:2855–2866

Daneholt B (2001) Assembly and transport of a premessenger RNP particle. Proc Natl Acad Sci USA 98:7012–7017

Dong B, Horowitz DS, Kobayashi R, Krainer AR (1993) Purification and cDNA cloning of HeLa cell p54nrb, a nuclear protein with two RNA recognition motifs and extensive homology to human splicing factor PSF and Drosophila NONA/BJ6. Nucleic Acids Res 21:4085–4092

Edström JE, Ericson E, Lindgren S, Lonn U, Rydlander L (1978) Fate of Balbiani-ring RNA in vivo. Cold Spring Harb Symp Quant Biol 42 Pt 2:877–884

Emili A, Shales M, McCracken S, Xie W, Tucker PW, Kobayashi R, Blencowe BJ, Ingles CJ (2002) Splicing and transcription-associated proteins PSF and p54nrb/nonO bind to the RNA polymerase II CTD. RNA 8:1102–1111

Fleckner J, Zhang M, Valcárcel J, Green MR (1997) U2AF65 recruits a novel human DEAD box protein required for the U2 snRNP-branchpoint interaction. Genes Dev 11:1864–1872

Fox AH, Lam YW, Leung AK, Lyon CE, Andersen J, Mann M, Lamond AI (2002) Paraspeckles: a novel nuclear domain. Curr Biol 12:13–25

Gatfield D, Le Hir H, Schmitt C, Braun IC, Kocher T, Wilm M, Izaurralde E (2001) The DExH/D box protein HEL/UAP56 is essential for mRNA nuclear export in Drosophila. Curr Biol 11:1716–1721

Gururajan R, Perry-O'Keefe H, Melton DA, Weeks DL (1991) The Xenopus localized messenger RNA An3 may encode an ATP-dependent RNA helicase. Nature 349:717–719

Hägele K (1975) Chironomus. In: King RC (ed) Handbook of genetics. Northwestern University, Evanston, IL, USA, pp 269–327

Hilleren P, McCarthy T, Rosbash M, Parker R, Jensen TH (2001) Quality control of mRNA 3′-end processing is linked to the nuclear exosome. Nature 413:538–542

Jensen TH, Boulay J, Rosbash M, Libri D (2001a) The DECD box putative ATPase Sub2p is an early mRNA export factor. Curr Biol 11:1711–1715

Jensen TH, Patricio K, McCarthy T, Rosbash M (2001b) A block to mRNA nuclear export in S. cerevisiae leads to hyperadenylation of transcripts that accumulate at the site of transcription. Mol Cell 7:887–898

Jones KR, Rubin GM (1990) Molecular analysis of no-on-transient A, a gene required for normal vision in Drosophila. Neuron 4:711–723

Kiesler E, Miralles F, Visa N (2002) HEL/UAP56 binds cotranscriptionally to the Balbiani ring pre-mRNA in an intron-independent manner and accompanies the BR mRNP to the nuclear pore. Curr Biol 12:859–862

Kiesler E, Miralles F, Östlund Farrants AK, Visa N (2003) The Hrp65 self-interaction domain is evolutionarily conserved and is required for nuclear import of Hrp65 isoforms that lack a nuclear localization signal. J Cell Sci 116:3949–3956

Kim VN, Dreyfus G (2001) Nuclear mRNA binding proteins couple pre-mRNA splicing and post-splicing events. Mol Cells 12:1–10

Kosova B, Pante N, Rollenhagen C, Podtelejnikov A, Mann M, Aebi U, Hurt E (2000) Mlp2p, a component of nuclear pore attached intranuclear filaments, associates with nic96p. J Biol Chem 275:343–350

Lei EP, Silver PA (2002) Protein and RNA export from the nucleus. Dev Cell 2:261–272

Leroy P, Alzari P, Sassoon D, Wolgemuth D, Fellous M (1989) The protein encoded by a murine male germ cell-specific transcript is a putative ATP-dependent RNA helicase. Cell 57:549–559

Linder P, Stutz F (2001) mRNA export: travelling with DEAD box proteins. Curr Biol 11:R961–R963

Lönnroth A, Alexciev K, Mehlin H, Wurtz T, Skoglund U, Daneholt B (1992) Demonstration of a 7-nm RNP fiber as the basic structural element in a premessenger RNP particle. Exp Cell Res 199:292–296

Luo ML, Zhou Z, Magni K, Christoforides C, Rappsilber J, Mann M, Reed R (2001) Pre-mRNA splicing and mRNA export linked by direct interactions between UAP56 and Aly. Nature 413:644–647

Mamiya N, Worman HJ (1999) Hepatitis C virus core protein binds to a DEAD box RNA helicase. J Biol Chem 274:15751–15756

Maniatis T, Reed R (2002) An extensive network of coupling among gene expression machines. Nature 416:499–506

Mathur M, Tucker PW, Samuels HH (2001) PSF is a novel corepressor that mediates its effect through Sin3A and the DNA binding domain of nuclear hormone receptors. Mol Cell Biol 21:2298–2311

Mehlin H, Daneholt B (1993) The Balbiani ring particle: a model for the assembly and export of RNPs from the nucleus? Trends Cell Biol 3:443–447

Mehlin H, Daneholt B, Skoglund U (1992) Translocation of a specific premessenger ribonucleoprotein particle through the nuclear pore studied with electron microscope tomography. Cell 69:605–613

Melcak I, Cermanova S, Jirsova K, Koberna K, Malinsky J, Raska I (2000) Nuclear pre-mRNA compartmentalization: trafficking of released transcripts to splicing factor reservoirs. Mol Biol Cell 11:497–510

Melcak I, Melcakova S, Kopsky V, Vecerova J, Raska I (2001) Prespliceosomal assembly on microinjected precursor mRNA takes place in nuclear speckles. Mol Biol Cell 12:393–406

Mintz PJ, Patterson SD, Neuwald AF, Spahr CS, Spector DL (1999) Purification and biochemical characterization of interchromatin granule clusters. EMBO J 18:4308–4320

Miralles F, Visa N (2001) Molecular characterization of Ct-hrp65: identification of two novel isoforms originated by alternative splicing. Exp Cell Res 264:284–295

Miralles F, Öfverstedt LG, Sabri N, Aissouni Y, Hellman U, Skoglund U, Visa N (2000) Electron tomography reveals posttranscriptional binding of pre-mRNPs to specific fibers in the nucleoplasm. J Cell Biol 148:271–282

Misteli T (2000) Cell biology of transcription and pre-mRNA splicing: nuclear architecture meets nuclear function. J Cell Sci 113(11):1841–1849

Misteli T, Cáceres JF, Spector DL (1997) The dynamics of a pre-mRNA splicing factor in living cells. Nature 387:523–527

Monneron A, Bernhard W (1969) Fine structural organization of the interphase nucleus in some mammalian cells. J Ultrastruct Res 27:266–288

Muhlemann O, Mock-Casagrande CS, Wang J, Li S, Custodio N, Carmo-Fonseca M, Wilkinson MF, Moore MJ (2001) Precursor RNAs harboring nonsense codons accumulate near the site of transcription. Mol Cell 8:33–43

Patton JG, Porro EB, Galceran J, Tempst P, Nadal-Ginard B (1993) Cloning and characterization of PSF, a novel pre-mRNA splicing factor. Genes Dev 7:393–406

Peng R, Dye BT, Perez I, Barnard DC, Thompson AB, Patton JG (2002) PSF and p54nrb bind a conserved stem in U5 snRNA. RNA 8:1334–1347

Percipalle P, Zhao J, Pope B, Weeds A, Lindberg U, Daneholt B (2001) Actin bound to hrp36 is associated with Balbiani ring mRNA from the gene to polysomes. J Cell Biol 153:229–236

Percipalle P, Fomproix N, Kylberg K, Miralles F, Björkroth B, Dameholt B, Visa N (2003) An actin-ribonucleoprotein interaction is involved in transcription by RNA polymerase II. Proc Natl Acad Sci USA 100:6475–6480

Politz JC, Pederson T (2000) Review: movement of mRNA from transcription site to nuclear pores. J Struct Biol 129:252–257

Politz JC, Browne ES, Wolf DE, Pederson T (1998) Intranuclear diffusion and hybridization state of oligonucleotides measured by fluorescence correlation spectroscopy in living cells. Proc Natl Acad Sci USA 95:6043–6048

Politz JC, Tuft RA, Pederson T, Singer RH (1999) Movement of nuclear poly(A) RNA throughout the interchromatin space in living cells. Curr Biol 9:285–291

Puvion E, Puvion-Dutilleul F (1996) Ultrastructure of the nucleus in relation to transcription and splicing: roles of perichromatin fibrils and interchromatin granules. Exp Cell Res 229:217–225

Reed R, Hurt E (2002) A conserved mRNA export machinery coupled to pre-mRNA splicing. Cell 108:523–531

Sabri N, Östlund Farrants AK, Hellman U, Visa N (2002) Evidence for a posttranscriptional role of a TFIIICalpha-like protein in Chironomus tentans. Mol Biol Cell 13:1765–1777

Schmitt C, von Kobbe C, Bachi A, Pante N, Rodrigues JP, Boscheron C, Rigaut G, Wilm M, Seraphin B, Carmo-Fonseca M, Izaurralde E (1999) Dbp5, a DEAD-box protein required for mRNA export, is recruited to the cytoplasmic fibrils of nuclear pore complex via a conserved interaction with CAN/Nup159p. EMBO J 18:4332–4347

Sewer MB, Nguyen VQ, Huang CJ, Tucker PW, Kagawa N, Waterman MR (2002a) Transcriptional activation of human CYP17 in H295R adrenocortical cells depends on complex formation among p54(nrb)/NonO, protein-associated splicing factor, and SF-1, a complex that also participates in repression of transcription. Endocrinology 143:1280–1290

Sewer MB, Waterman MR (2002b) Adrenocorticotropin/cyclic adenosine 3′,5′-monophosphate-mediated transcription of the human CYP17 gene in the adrenal cortex is dependent on phosphatase activity. Endocrinology 143:1769–1777

Shatkin AJ, Manley JL (2000) The ends of the affair: capping and polyadenylation. Nat Struct Biol 7:838–842

Shav-Tal Y, Zipori D (2002) PSF and p54(nrb)/NonO- multi-functional nuclear proteins. FEBS Lett 531:109–114

Singh OP, Björkroth B, Masich S, Wieslander L, Daneholt B (1999) The intranuclear movement of Balbiani ring premessenger ribonucleoprotein particles. Exp Cell Res 251:135–146

Skoglund U, Andersson K, Björkroth B, Lamb MM, Daneholt B (1983) Visualization of the formation and transport of a specific hnRNP particle. Cell 34:847–855

Skoglund U, Andersson K, Strandberg B, Daneholt B (1986) Three-dimensional structure of a specific pre-messenger RNP particle established by electron microscope tomography. Nature 319:560–564

Snay-Hodge CA, Colot HV, Goldstein AL, Cole CN (1998) Dbp5p/Rat8p is a yeast nuclear pore-associated DEAD-box protein essential for RNA export. EMBO J 17:2663–2676

Spector DL (1996) Nuclear organization and gene expression. Exp Cell Res 229:189–197

Strahm Y, Fahrenkrog B, Zenklusen D, Rychner E, Kantor J, Rosbach M, Stutz F (1999) The RNA export factor Gle1p is located on the cytoplasmic fibrils of the NPC and physically interacts with the FG-nucleoporin Rip1p, the DEAD-box protein Rat8p/Dbp5p and a new protein Ymr 255p. EMBO J 18:5761–5777

Strässer K, Hurt E (2001) Splicing factor Sub2p is required for nuclear mRNA export through its interaction with Yra1p. Nature 413:648–652

Sun X, Alzhanova-Ericsson AT, Visa N, Aissouni Y, Zhao J, Daneholt B (1998) The hrp23 protein in the Balbiani ring pre-mRNP particles is released just before or at the binding of the particles to the nuclear pore complex. J Cell Biol 142:1181–1193

Tanner NK, Linder P (2001) DExD/H box RNA helicases: from generic motors to specific dissociation functions. Mol Cell 8:251–262

Tseng SS, Weaver PL, Liu Y, Hitomi M, Tartakoff AM, Chang TH (1998) Dbp5p, a cytosolic RNA helicase, is required for poly(A)+ RNA export. EMBO J 17:2651–2662

Visa N, Puvion-Dutilleul F, Harper F, Bachellerie JP, Puvion E (1993) Intranuclear distribution of poly(A) RNA determined by electron microscope in situ hybridization. Exp Cell Res 208:19–34

Visa N, Izaurralde E, Ferreira J, Daneholt B, Mattaj IW (1996a) A nuclear cap-binding complex binds Balbiani ring pre-mRNA cotranscriptionally and accompanies the ribonucleoprotein particle during nuclear export. J Cell Biol 133:5–14

Visa N, Alzhanova-Ericsson AT, Sun X, Kiseleva E, Bjorkroth B, Wurtz T, Daneholt B (1996b) A pre-mRNA-binding protein accompanies the RNA from the gene through the nuclear pores and into polysomes. Cell 84:253–264

Wetterberg I, Zhao J, Masich S, Wieslander L, Skoglund U (2001) In situ transcription and splicing in the Balbiani ring 3 gene. EMBO J 20:2564–2574

Wieslander L (1994) The Balbiani ring multigene family: coding repetitive sequences and evolution of a tissue-specific cell function. Prog Nucleic Acid Res Mol Biol 48:275–313

Will CL, Lührmann R (2001a) Spliceosomal UsnRNP biogenesis, structure and function. Curr Opin Cell Biol 13:290–301

Will CL, Lührmann R (2001b) Molecular biology. RNP remodeling with DExH/D boxes. Science 291:1916–1917

Xing Y, Johnson CV, Moen PT Jr, McNeil JA, Lawrence J (1995) Nonrandom gene organization: structural arrangements of specific pre-mRNA transcription and splicing with SC-35 domains. J Cell Biol 131:1635–1647

Yang YS, Yang MC, Tucker PW, Capra JD (1997) NonO enhances the association of many DNA-binding proteins to their targets. Nucleic Acids Res 25:2284–2292

Zhang Z, Carmichael GG (2001) The fate of dsRNA in the nucleus: a p54(nrb)-containing complex mediates the nuclear retention of promiscuously A-to-I edited RNAs. Cell 106:465–475

Zhao J, Jin SB, Bjorkroth B, Wieslander L, Daneholt B (2002) The mRNA export factor Dbp5 is associated with Balbiani ring mRNP from gene to cytoplasm. EMBO J 21:1177–1187

Zimowska G, Paddy MR (2002) Structures and dynamics of Drosophila Tpr inconsistent with a static, filamentous structure. Exp Cell Res 276:223–232

Photobleaching Microscopy Reveals the Dynamics of mRNA-Binding Proteins Inside Live Cell Nuclei

José Braga[1], José Rino[1] and Maria Carmo-Fonseca[1]

1
Introduction

Immediately upon synthesis nascent transcripts associate with proteins to form large ribonucleoprotein (RNP) complexes, the protein content of which evolves as pre-mRNA is processed into mRNA and nuclear mRNA is exported to the cytoplasm (for recent reviews, see Dreyfuss et al. 2002; Reed and Hurt 2002). After being released from the gene templates, which are distributed throughout the nucleoplasm, messenger ribonucleoprotein particles (mRNPs) must reach the nuclear pore complexes in order to be transported to the cytoplasm – but what drives this intranuclear movement remains unknown.

Recently, we have developed quantitative photobleaching methods to investigate the mobility of mRNPs within the nucleus of living human cells (Calapez et al. 2002). RNP complexes containing mRNA were made fluorescent by transient expression of GFP fused to two distinct mRNA-binding proteins, PABP2/PABPN1 and TAP/NXF1.

The nuclear poly(A)-binding protein (PABP2 or PABPN1) binds to the growing poly(A) tails formed at the 3′-ends of nearly all eukaryotic mRNAs (Wahle 1991). PABP2 cooperates with the cleavage and polyadenylation specificity factor (CPSF) to stimulate the activity of poly(A) polymerase, the enzyme that catalyses polyadenylation. The cleavage and polyadenylation reactions are currently thought to be coupled to splicing of the last intron and to occur at the same time as or just before transcription termination (Bauren et al. 1998; Proudfoot et al. 2002). This implies that GFP-PABP2 will bind to nearly terminated and spliced transcripts.

In addition to GFP-PABP2, we aimed at achieving a more specific visualization of mRNAs in transit to the cytoplasm by using GFP fused to the export factor TAP (also called NXF1). TAP binds directly to the constitutive transport element (CTE) of viral RNAs and is required for the nucleo-cytoplasmic export of cellular mRNAs in vertebrates, *Caenorhabditis elegans*, and *Saccharomyces cerevisiae* (reviewed in Görlich and Kutay 1999; Conti and Izaurralde 2001). TAP associates with cellular mRNPs and is thought to promote their export by interacting with nuclear pore proteins during translocation (Bachi et al. 2000).

[1]Institute of Molecular Medicine, Faculty of Medicine, University of Lisbon, 1649-028, Lisbon, Portugal. Tel. +351-21-7940157; Fax. +351-21-7951780, e-mail: carmo.fonseca@fm.ul.pt

Progress in Molecular and Subcellular Biology
P. Jeanteur (Ed.): RNA Trafficking and Nuclear Structure Dynamics
© Springer-Verlag Berlin Heidelberg 2003

In photobleaching experiments, a small region in the nucleus is illuminated at high intensity by a high-powered focused laser beam. As a consequence, GFP molecules present within that region are irreversibly bleached without detectably damaging the cell. Subsequent diffusion of bleached GFP molecules out of the bleached area and diffusion of surrounding nonbleached GFP molecules into the bleached area leads to a recovery of fluorescence, which is recorded at low laser power. There are two variations of photobleaching experiments that yield different types of information: fluorescence recovery after photobleaching (FRAP), and fluorescence loss in photobleaching (FLIP). In FRAP, a region is bleached once and subsequent recovery of fluorescence in the bleached area is monitored over time. In FLIP, a region is repeatedly bleached, and the loss of fluorescence from the surrounding nonbleached area is monitored over time (reviewed in White and Stelzer 1999; Reits and Neefjes 2001).

2
General Concepts of Quantitative FRAP

FRAP microscopy is currently used in several laboratories to study the mobility of GFP chimeras in the living cell nucleus (Ellenberg and Lippincott-Schwartz 1999; Houtsmuller and Vermeulen 2001; Phair and Misteli 2001). Assuming that macromolecular mobility within the nucleus is primarily caused by Brownian motion (reviewed in Carmo-Fonseca et al. 2002), two parameters can be estimated from FRAP: the fraction of GFP fusion protein that is mobile and its rate of mobility expressed as the diffusion coefficient, D (White and Stelzer 1999; Reits and Neefjes 2001). D indicates the surface area randomly sampled by the fluorescent molecules in a given time (usually $\mu m^2 s^{-1}$).

To estimate a diffusion coefficient, the recovery of relative fluorescence intensity within the bleach region is plotted as a function of time. Typically, each FRAP analysis starts with three image scans, followed by a single bleach pulse of 37 ms on a spot with a diameter of 25 pixels (0.71-μm-radius). A series of 97 single-section images are then collected at 78-ms intervals, with the first image acquired 2 ms after the end of bleaching. Image size is 512×50 pixels and the pixel width is 57 nm. For imaging, the laser power is attenuated to 0.1–0.2% of the bleach intensity. For each experiment, the background and nuclear regions can be automatically identified using a segmentation algorithm (Fig. 1A, B). The average fluorescence in the nucleus $T(t)$ and the average fluorescence in the bleached region $I(t)$ are then calculated for each background subtracted image at time t after bleaching, and the recovery curves are normalized according to the equation reported by Phair and Misteli (2000):

$$Irel(t) = \frac{I(t)}{I_i} \frac{T_i}{T(t)} \tag{1}$$

where T_i is the fluorescence in the nucleus before bleaching and I_i is the fluorescence in the bleached region before bleaching. This normalization corrects for the loss of fluorescence caused by imaging. Typically, ~10% of the total GFP fluorescence is lost during the bleach pulse, whereas during the post-bleaching scanning phase the fluorescence lost is <5% (Calapez et al. 2002).

D is calculated by fitting a function, $Irel^*(t)$, to the fluorescence recovery curve (Fig. 1C, D). This function is derived from a theoretical model originally developed by Axelrod et al. (1976). Photobleaching theory assumes that fluorescent molecules can be either mobile (with a diffusion coefficient D) or immobile. The initial concentration of fluorescent molecules inside the bleached zone is C_0, γ being the percentage of immobile molecules (ranging from 0 to 1). Total fluorescence observed inside the bleached spot is then the sum of the fluorescence of the mobile molecules and the fluorescence of the immobile ones. After bleaching, the immobile fraction fluorescence is constant inside the bleached spot:

$$F_{im}(t) = \kappa C_0 \gamma \frac{1 - e^{-K}}{K} \tag{2}$$

where K is the bleach constant, which is a measure of the intensity of the bleaching laser and the properties of the fluorophore and κ is a parameter that depends on the properties of the laser and of the detection system. The mobile fraction fluorescence is given by:

$$F_m(t) = \kappa C_0 (1 - \gamma) \sum_{n=0}^{+\infty} \frac{(-K)^n}{n!} \left(1 + n\left(1 + 2\frac{t}{\tau_D} \right) \right)^{-1} \tag{3}$$

where τ_D is the characteristic time of diffusion and is related with the diffusion coefficient by $D = \frac{w^2}{4\tau_D}$. The total normalized fluorescence is then:

$$Irel * (t) = (1 - \gamma) \sum_{n=0}^{+\infty} \frac{(-K)^n}{n!} \left(1 + n\left(1 + 2\frac{t}{\tau_D} \right) \right)^{-1} + \gamma \frac{1 - e^{-K}}{K} \tag{4}$$

According to this equation, the normalized fluorescence corresponds to the weighted-sum of the fluorescence recovery of the mobile portion of the fluorophore (first term) and the fluorescence after bleaching of the immobile portion of the fluorophore (second term: see also Klonis et al. 2002).

For a laser with a Gaussian intensity profile, the theoretically predicted post-bleach concentration profile of immobile molecules is:

$$C(r) = C_0 \exp\left(-k \exp\left(-2\frac{r^2}{w^2} \right) \right) \tag{5}$$

where k is the bleach constant for immobile molecules and w is half of the width of the laser intensity at $1/e^2$ intensity. Phair and Misteli (2000) used chemically fixed cells to fit Eq. 5 to the observed concentration profile of fluorescent molecules after the bleach, obtaining estimates for k and w. The value of k was then used as the bleach constant K in Eq. 4. This approach is valid

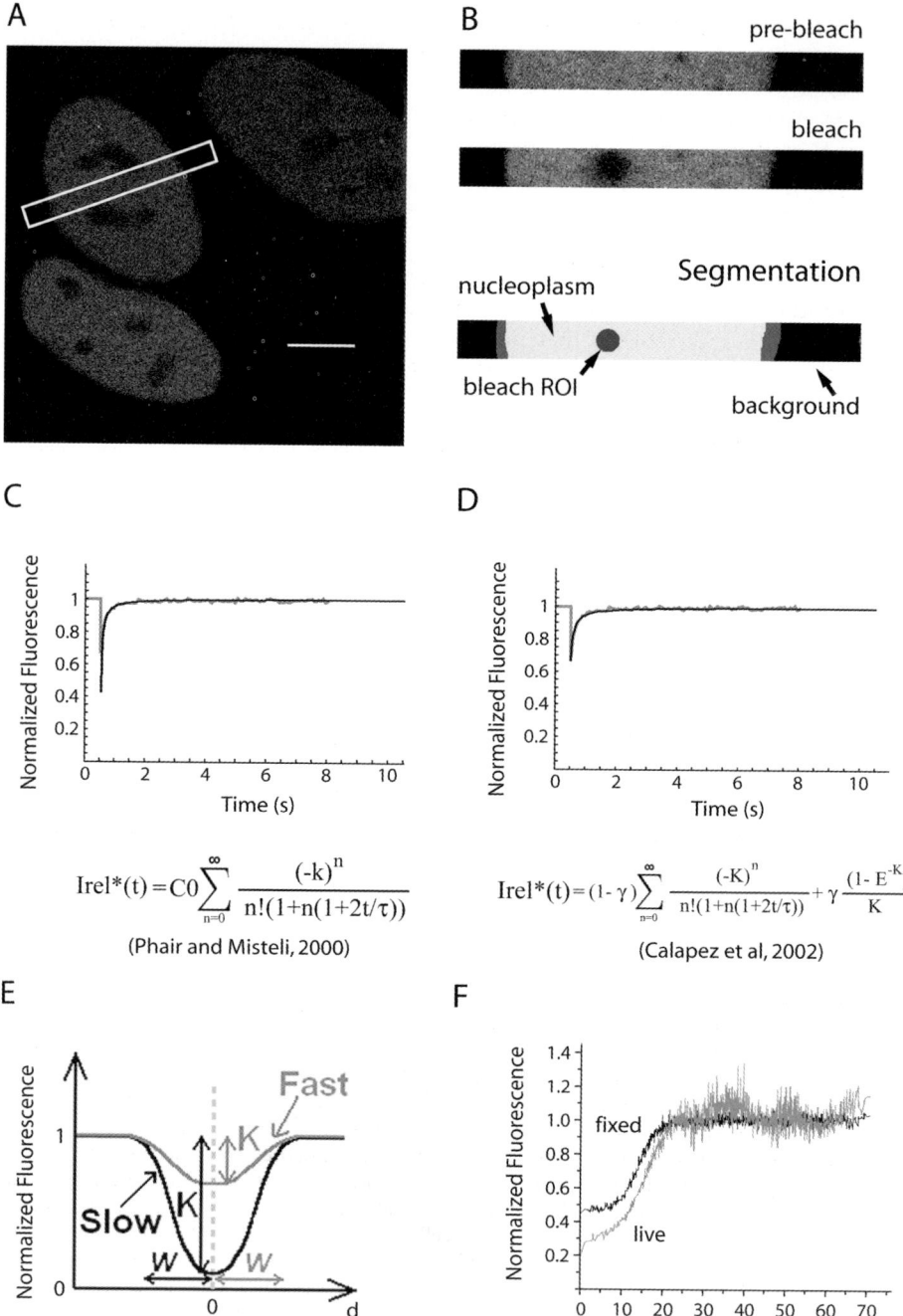

A

B pre-bleach

bleach

nucleoplasm   Segmentation

bleach ROI

background

C

$$Irel*(t) = C0 \sum_{n=0}^{\infty} \frac{(-k)^n}{n!(1+n(1+2t/\tau))}$$

(Phair and Misteli, 2000)

D

$$Irel*(t) = (1-\gamma) \sum_{n=0}^{\infty} \frac{(-K)^n}{n!(1+n(1+2t/\tau))} + \gamma \frac{(1-E^{-K})}{K}$$

(Calapez et al, 2002)

E Fast

K

Slow   K

W   W

Distance to spot center

F fixed

live

Distance to spot center (pixel)

assuming that no significant diffusion occurs during bleaching. However, in the case of molecules that diffuse at fast rates, the values of k and w are significantly different from those estimated in fixed cells (Fig. 1E). We therefore proposed an alternative method, in which K is an additional free parameter to be estimated from the FRAP recovery curve fit to Eq. 4 (Calapez et al. 2002). According to this method, Eq. 5 is used to calculate the value of w, but the value of k is disregarded. When compared to the method used by Phair and Misteli (Fig. 1C), the approach proposed in Calapez et al. (2002) results in a better fit (Fig. 1D), particularly in the case of highly mobile molecules.

Although it has been common practice in FRAP studies to determine the bleaching characteristics of the laser by bleaching a spot in a fixed specimen, we observed that chemical fixation alters the post-bleach concentration profile of GFP (Fig. 1F). This implies it is more accurate to estimate w by bleaching a GFP fusion protein that is immobile in the living cell rather than by bleaching a fixed specimen.

Fig. 1. **A** HeLa cells expressing GFP fused to a point mutant variant of PABP2 (GFP-PABP2dm) with impaired ability to bind to poly(A)RNA. This protein distributes homogeneously throughout the nucleoplasm with lower concentration in nucleoli (Calado and Carmo-Fonseca 2000). The *white box* indicates the region scanned in a FRAP experiment. *Bar* 5 μm. **B** Pre- and postbleach images of the scanned region depicted in **A**. These images were segmented for quantification of fluorescence intensity. **C, D** A plot of fluorescence intensity (*I*) as a function of time shows the parameters of a quantitative FRAP experiment. The recovery curve depicted (*grey line*) corresponds to a pool of four independent experiments, with ten different cells analyzed per experiment. The bleach region is imaged during a pre-bleach period to determine the initial intensity I_i. This region is then bleached using high-intensity laser power. As a result, the fluorescence intensity decreases from the initial value I_i to I_0. Recovery is monitored starting at time t_0 until I reaches a final value or plateau, when no further increase can be detected (I_∞). I is corrected for the background intensity and the amount of total fluorescence lost during the bleach and imaging. The experimental FRAP recovery curve is then fitted to the theoretical recovery function, $Irel^*(t)$ (*black line*). The parameter k corresponds to the bleach constant of immobilized molecules in fixed cells, whereas K corresponds to the bleach constant of mobile molecules in live cells. Note that the equation used in **D** results in a better fit of the theoretical model to the experimental data. **E** The concentration profile of fluorescent molecules after bleaching (i.e., the post-bleaching concentration of fluorophore at distance d from the center of the bleached region) is distinct for molecules that move at either slow or fast rates. The parameter K (bleach constant of mobile molecules in live cells) depends on the bleaching laser and on the properties of the fluorophore; w is the half width of the laser beam at $1/e^2$ intensity. The concentration profile of a fluorophore after bleaching live cells is similar to that obtained in fixed cells only if molecules are largely immobile or diffuse very slowly. **F** Chemical fixation interferes with the bleaching properties of GFP. A chimera formed by fusion of PABP2 with a N-terminal fragment of p80 coilin (amino acids 1–468) is immobile in the nucleus (Calapez et al. 2002). Distinct GFP concentration profiles are obtained when cells are bleached either alive or after fixation. Due to the radial symmetry of the profile, only positive values of d are shown

3
FRAP Reveals the Dynamics of GFP-PABP2 and GFP-TAP in the Living Cell Nucleus

After establishing that PABP2 and TAP GFP fusions behave similarly to endogenous proteins (Bachi et al. 2000; Calado et al. 2000; Calapez et al. 2002), FRAP experiments were performed on transiently transfected HeLa cells. To monitor expression level of GFP-fusion proteins, the fluorescence intensity of each transfected nucleus was determined using the same detection parameters as those applied during the imaging stage of photobleaching experiments. Cells containing the minimal detectable level of GFP-fusion proteins were selected in order to avoid potential artifacts caused by over-expression.

For comparison purposes, we analysed cells expressing GFP-PABP2, GFP-TAP and two GFP fusions to mutant forms of PABP2 and TAP that fail to bind to mRNA (GFP-PABP2dm and GFP-TAP371-619, respectively). The variant PABP2dm contains a double-point mutation (a tyrosine to alanine at position 175 and a phenylalanine to alanine at position 215) in the RNA binding domain of PABP2. In vitro this mutant binds to poly(A) approximately 20-fold less than wild-type PABP2 (Calado et al. 2000). The TAP fragment comprising residues 371–619 excludes both the RNA binding domain and the leucine-rich repeats (Bachi et al. 2000), which play an essential role in TAP-mediated export of mRNA (Braun et al. 2001).

In cells expressing the mutant GFP fusions, the fluorescence signal recovers to the pre-bleach value in less than 1 s (Fig. 2A, B). In contrast, GFP-PABP2 requires >7 s to recover completely (Fig. 2C) and GFP-TAP recovers to the pre-bleach value in 4 s (Fig. 2D). The calculated D values for GFP-PABP2 and GFP-TAP are 0.6 and 1.2 $\mu m^2 s^{-1}$, respectively, whereas for the mutants D is 4.2 $\mu m^2 s^{-1}$ and 4.3 $\mu m^2 s^{-1}$. Thus, the kinetics of wild-type GFP-fusion molecules is approximately four to sevenfold slower than mutant proteins. In the particular case of PABP2 fusion proteins, both the wild-type and mutant forms have the same molecular size, but differ in their ability to bind to poly(A) RNA. Thus, the retardation of GFP-PABP2 molecules is most likely a direct consequence of binding to mRNP complexes.

The GFP fusions to PABP2dm (which contains 306 amino acids) and TAP deletion mutant (a 238-amino acid protein) diffuse in the nucleus with similar kinetics (4.2 and 4.3 $\mu m^2 s^{-1}$, respectively). We assume that these mutants with impaired ability to bind to RNA diffuse in the nucleoplasm essentially as free proteins. In contrast, the wild-type proteins are slowed down because they form large mRNP complexes. Intriguingly, GFP-PABP2 and GFP-TAP diffuse at significantly distinct rates. This could indicate that the two fusion proteins are tagging distinct populations of mRNPs with different mobility rates. However, before concluding that the diffusion coefficients estimated for GFP-PABP2 and GFP-TAP represent a measure of the mobility rate of mRNP complexes in the nucleus, it is crucial to determine the fraction of GFP fusion molecules that

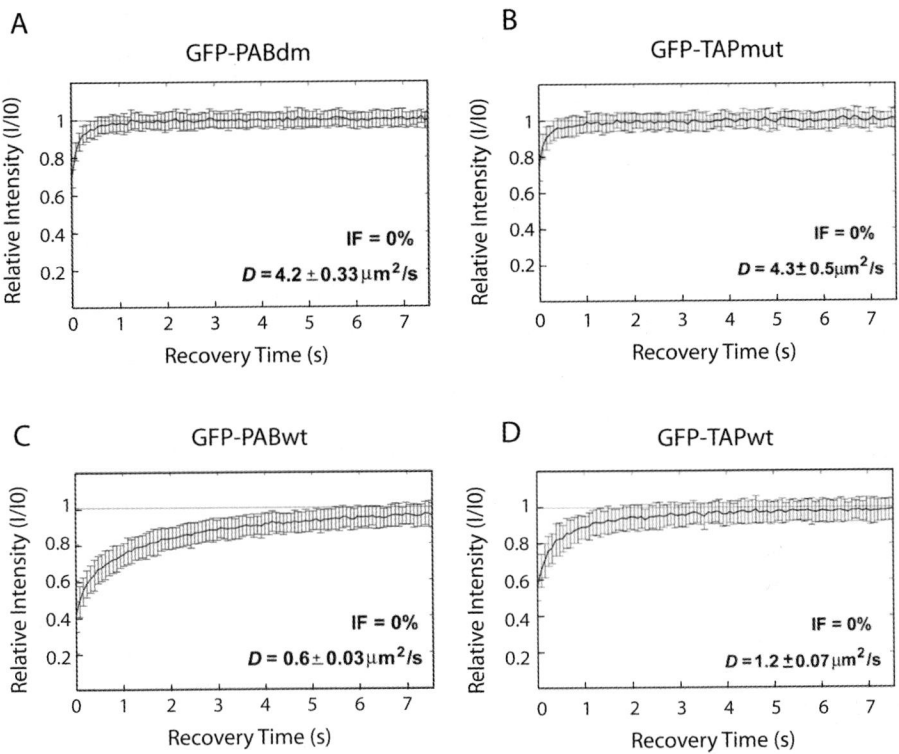

Fig. 2. HeLa cells expressing either mutant forms of PABP2 (GFP-PABP2dm, **A**), and TAP (GFP-TAPmut, **B**) or the wild-type proteins (GFP-PABP2wt, **C**; and GFP-TAPwt, **D**) were imaged before and during recovery after bleaching. The recovery curves correspond to a pool of three to five independent experiments, with ten different cells analyzed per experiment. *Error bars* represent standard deviations. *D* values represent mean ± standard error

is actually bound to mRNP complexes. Although fitting of FRAP recovery curves to an exponential function was used to characterized fast and slow populations of a GFP chimera to RNA polymerase II (Kimura et al. 2002), the simulation presented in Fig. 3A shows that the recovery of a freely mobile molecule diffusing at $1.0\ \mu m^2 s^{-1}$ is slower than expected for an exponential function. In fact, an exponential function applies when there is a linear exchange of molecules between two compartments. In a FRAP experiment, neither the bleached nor the unbleached regions can be considered as compartments because the concentration of fluorescent molecules within each region is not uniform due to diffusion during the bleach. The simulations presented in Fig. 3B and C further demonstrate that Axelrod's model for quantitative FRAP does not resolve a mixed population of GFP fusion proteins such as that composed by a fraction of slow-moving molecules bound to mRNP complexes and a fast-moving fraction of free molecules. In a system containing a mix of two mobile populations diffusing at different rates, a single diffusion coefficient is estimated by FRAP. In general, this will be lower than the weighted mean of the individual diffusion coefficients of each population (Fig. 3D). Thus, there is not a simple relationship between the individual diffusion coefficients and the estimated diffusion coefficient for the mixture. To discriminate between diffusion rates of mRNP-bound and free GFP fusion molecules, it is necessary to perform a variant form of photobleaching analysis, FLIP.

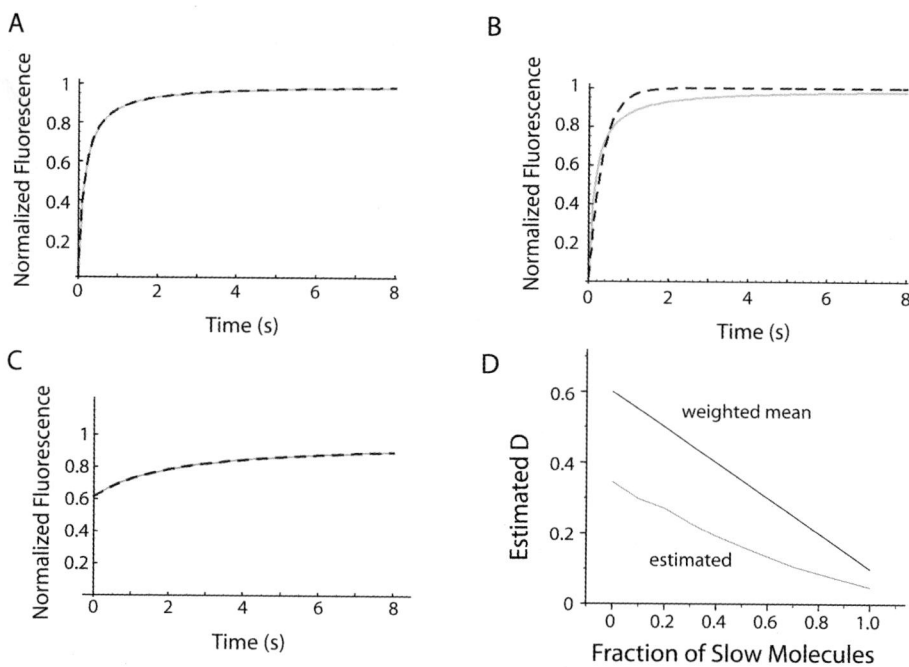

Fig. 3. **A, B** A FRAP recovery curve was simulated for a single population of freely diffusing molecules (*grey line*) and fitted to a theoretical recovery function (*black dashed line*). **A** depicts the fit to Axelrod's model. **B** shows the fit to an exponential function. The data clearly indicate that the recovery curve is not exponential. **C** A FRAP recovery curve was simulated for a mixed population of freely diffusing molecules (*grey line*). We have considered two molecular species moving at different rates ($D_1 = 0.1 \ \mu m^2 s^{-1}$ and $D_2 = 0.6 \ \mu m^2 s^{-1}$) present in equal amounts. Axelrod's model, which does not take into account fast and slow diffusing molecules in a mixed population, results in a very good fit to the recovery curve (*black dashed line*). **D** FRAP recovery curves were simulated for a mixed population of freely diffusing molecules. Two molecular species moving at different rates were considered to be present in different proportions. The graph depicts the diffusion coefficient values deduced from the simulated FRAP recovery curves (*grey line*) and the weighted means of the "real" diffusion coefficient from each molecular species (*black line*). Clearly, the deduced diffusion coefficients tend to be lower than the corresponding weighted means. Note that Axelrod's fitting method has a tendency to underestimate the "real" diffusion coefficients

4
Estimating the Fraction of GFP Fusion Proteins Bound to mRNP Complexes by FLIP

To perform FLIP, a region in the nucleus is repeatedly bleached using high laser intensity while the surrounding area is imaged between each round of bleaching. The loss of fluorescence in the area outside the bleached region is plotted over time, providing information on the rate of displacement of the molecules from that particular space. In the theoretical model proposed by Phair and Misteli (2001), diffusion during bleaching can be neglected (i.e., the free or unbound protein moves so fast in the time-scale of the FLIP experiment that it can be treated as being well-mixed in the nuclear compartment). However, when performing FLIP experiments using GFP fusions to PABP2 and TAP we observe that the concentration of fluorophore is not the same throughout the nucleus, but rather decreases as the bleach area is approached. Thus, diffusion cannot be neglected and the compartmental modelling approach is not valid. In this case, FLIP experiments can be used to resolve populations of molecules that are diffusing at different rates.

Simulation of a FLIP experiment with freely mobile molecules diffusing at $0.1\ \mu m^2\,s^{-1}$ shows that the decline in fluorescence is mainly due to diffusion during each bleach period (Fig. 4A, B). If there is a single population of GFP fusion molecules diffusing in the nucleoplasm, the plot of fluorescence loss will be an exponential (a straight line when plotted on semi-log graphs, Fig 4C). However, if there are two populations, one diffusing faster than the other, loss of fluorescence will be the sum of two exponentials (Fig. 4D, E). Fitting the FLIP curve to the nonlinear function of two-exponential decay

$$F(t) = p\exp(-R_1 t) + (1-p)\exp(-R_2 t) \tag{6}$$

yields the proportion of fast and slow populations, as well as the rate constants associated with the loss of fluorescence (Calapez et al. 2002). Additional slow phases in FLIP curves may appear due to the import of unbleached molecules from the cytoplasm or release from an intranuclear-bound pool of protein, if these phenomena occur in the time scale of the FLIP experiment.

Kinetic analysis of FLIP data for cells expressing low levels of GFP-PABP2 revealed the existence of a single slow-moving population, indicating that the vast majority of the fusion protein is bound to RNA (Calapez et al. 2002). In cells expressing higher levels of GFP-PABP2, the analysis revealed two populations, one slow (with kinetics similar to that detected in low-level-expressing cells) and one faster (with kinetics similar to the RNA-binding defective mutant). This suggests that the faster population of GFP fusion molecules represents a pool of free proteins unbound to RNA. In contrast to the results obtained with GFP-PABP2, FLIP analysis of cells expressing low levels of GFP-TAP shows the presence of a mixed population of slow and fast moving molecules suggesting that approximately 44% of TAP molecules in the nucleus are not bound to mRNP complexes (J. Rino et al., manuscript in prep.).

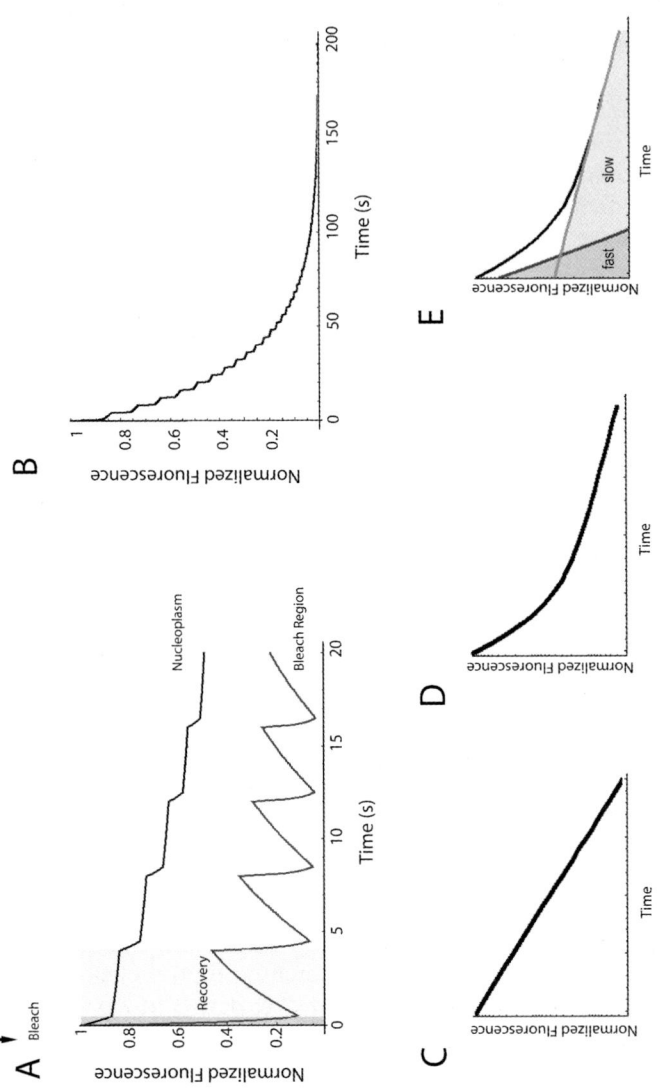

Fig. 4. A, B, C A FLIP experiment was simulated for a single population of freely diffusing molecules ($D = 0.1\ \mu m^2\ s^{-1}$). During each bleach there is a fluorescence loss both in the bleached region (**A** *thick line*) and in the unbleached nucleoplasm (**A** *thin line*). After each bleach (**A** *bleach*), the fluorescence increases in the bleached region (**A** *thick line*) while it continues to decrease in the unbleached nucleoplasm (**A** *thin line*). Note that fluorescence loss in the unbleached region is higher during the bleach period than during recovery. **B** shows the fluorescence loss in the unbleached nucleoplasm after prolonged repetitive bleaching. **C** shows the same data plotted on a semi-log graph. **D, E** A FLIP experiment was simulated for a mixed population of freely diffusing molecules. We have considered a fraction of slow-moving molecules (representing 30% of the total population) and a fraction of faster molecules moving at a rate ten times higher. **D** shows a semi-log plot of global fluorescence loss in the unbleached region. **E** depicts separately the fluorescence loss for each population of molecules present in the mix

In conclusion the data indicate that, when expressed at a low level, the vast majority of GFP-PABP2 molecules are incorporated into RNP complexes. Consequently, the diffusion coefficient estimated by FRAP represents a measure of the mobility rate of mRNPs. Under the same conditions, approximately half the GFP-TAP molecules roam the nucleus as free protein. Because FRAP experiments do not resolve individual mobility rates in a mixed population of molecules (i.e., RNA-bound and free GFP-TAP), the estimated diffusion coefficient for GFP-TAP is not representative of the mobility rate of mRNPs.

5
Simulation of mRNP Mobility Within the Nucleus Using a Random Walk Model

Based on the results described above, we assume that mRNPs move within the nucleus at a rate of $0.6 \pm 0.03\ \mu m^2\,s^{-1}$. A very similar value, $0.6 \pm 0.1\ \mu m^2\,s^{-1}$, was reported by Politz et al. (1999) using a distinct experimental approach. In this study, nuclear poly(A) RNA was visualized in living cells by hybridization with an oligo(dT) probe labelled with chemically masked (caged) fluorescein. Upon unmasking the fluorescence by laser spot photolysis, the spread of the signal was followed by time-lapse microscopy (Politz et al. 1999).

In face of the limited experimental data available on this topic, numerical models have the potential of providing an important means of investigating the dynamics of mRNPs in the living cell nucleus. This prompted us to simulate the behavior of an mRNP particle assuming that it undergoes a 3D Pearson-type random walk inside the volume of a sphere, which represents the cell nucleus (Fig. 5A). The model assumes that the particles do not interact with each other and cannot enter inside the volume occupied by nucleoli (represented by smaller spheres in the interior of the nucleus). Following a random path, the particles will eventually reach a nuclear pore (represented by circular holes randomly scattered at the surface of the nuclear sphere). According to Pearson's random walk model, a particle moves along a path composed of a number of steps and each step has a fixed length (l) with a randomly and isotropically chosen direction (Martins et al. 2000). The relation between the time (τ) and the number N of steps taken is (for N sufficiently large):

$$\tau = N \frac{l^2}{6D} \tag{7}$$

where D is the diffusion coefficient of mRNPs ($0.6\ \mu m^2\,s^{-1}$). This algorithm properly simulates Brownian motion, as validated by the data presented in Fig. 5B, C.

In order to obtain a value for l (which should be as small as possible), simulations were performed varying this parameter. The results show that the mean time required for a particle to reach the pore from a central position

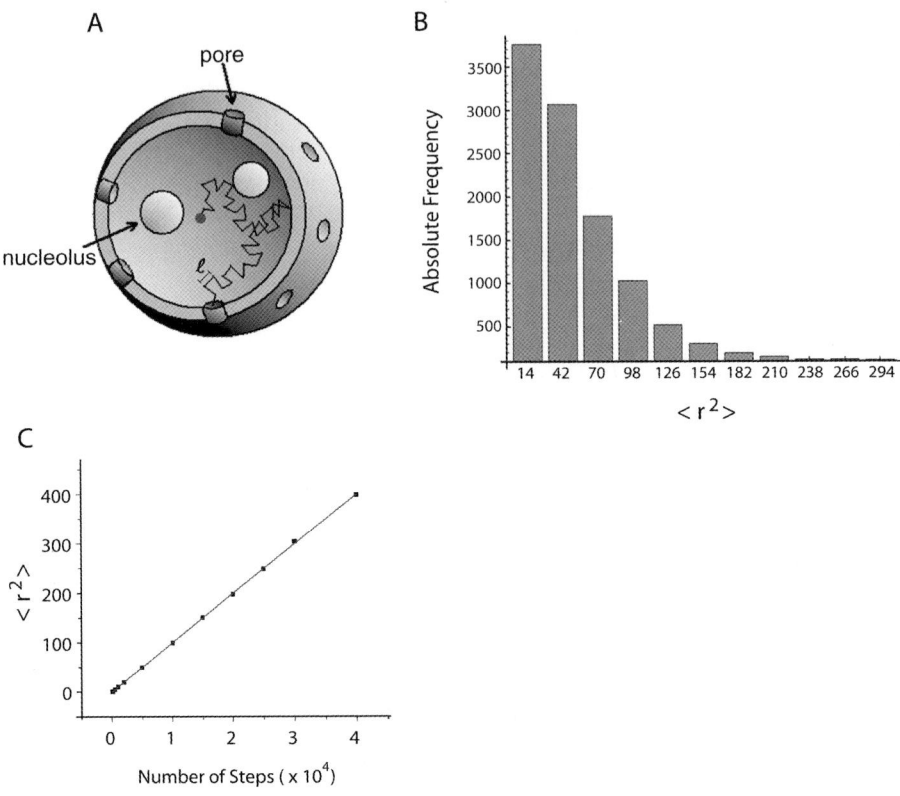

Fig. 5. A The cell nucleus is represented by a sphere of radius R. The nuclear pore complexes are represented by circular holes randomly scattered at the sphere's surface. Nucleoli are represented by smaller spheres contained in the interior of the nucleus. An mRNP particle is synthesized at different distances from the nuclear pores, depending on the localization of its cognate gene template. The particle will make a 3D Pearson-type random walk inside the nucleus, but it cannot enter nucleoli. Each random walk step has a fixed length, l. After a number of steps N, the particle will reach a nuclear pore. The relation between time and the number of steps is: $\tau = N\dfrac{l^2}{6D}$. **B, C** To validate that this algorithm properly describes Brownian motion, the evolution of a large ensemble of particles moving without any constraints was simulated. **B** shows a histogram obtained with 1×10^4 random walking particles moving 2×10^4 steps with length 0.1 (arbitrary units), showing the expected exponential decay. **C** shows that the dependency of $\langle r^2 \rangle$ with number of steps N agrees with the random walk model: $\langle r^2 \rangle = l^2 N$

does not significantly differ for values of l ranging between 1 and 4×10^{-9} m (Fig. 6A). To minimize the computation time (which increases significantly as the values of l decrease), a fixed value of 4 nm was chosen.

The nuclei of HeLa cells are ellipsoid with average radii ranging between 5 and 8 μm (Ribbeck and Görlich 2001). The average nuclear pore density in

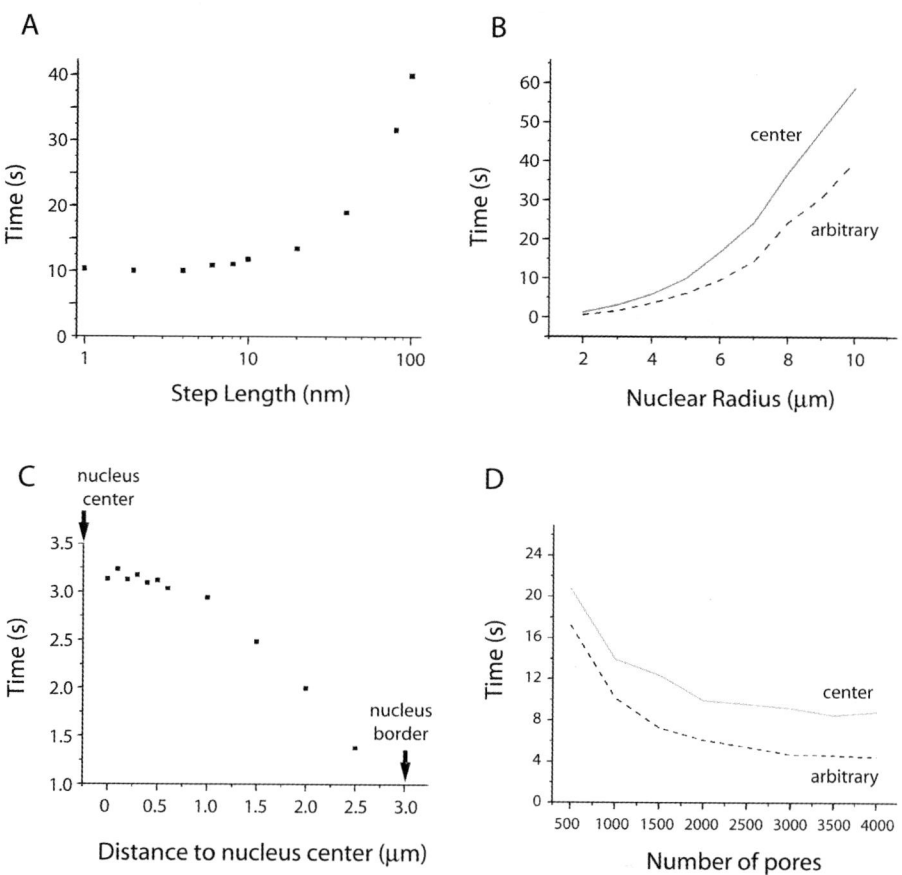

Fig. 6. A Time dependence on step length, showing that the mean time taken by the particle to reach the pore does not depend significantly on step length in the range between 1 and 4×10^{-9} m. **B** Time dependence on nuclear radius, for particles with an arbitrary initial position (*dashed line*) or departing from the center of the nucleus (*black line*). The number of pores was set to 2000 while the pore diameter was 40 nm. **C** Time dependence on initial position of the particle. The simulated nucleus had 2000 pores of 40-nm diameter and a radius of 3.0 µm. **D** Time dependence on the number of pores for particles with an arbitrary initial position (*dashed line*) or departing from the center of the nucleus (*black line*). The pores were 40-nm diameter and the cell nucleus was 5.0 µm

HeLa cells is 5 µm^{-2} (Kubitscheck et al. 1996; Ribbeck and Görlich 2001) corresponding to a total number of pores per nucleus of approximately 2770 (Ribbeck and Görlich 2001). The functional diameter of the nuclear pore complex has been determined using nucleoplasmin-coated gold and the largest particles that could still pass had a diameter of 25 nm (Feldherr et al. 1984). Since nucleoplasmin has to bind importins before nuclear pore complex passage (Weis 2002), it was proposed that the nuclear pore channel has a total diameter of ~40 nm (Ribbeck and Görlich 2001).

Assuming a nucleus with 2000 pores and a pore diameter of 40 nm, simulations were performed to determine the time required for an mRNP to reach a nuclear pore. As shown in Fig. 6B, the time varies depending on the radius of the nucleus. For example, an mRNA transcribed from a gene located in the center of the nucleus will take 1.3 s to reach a pore if the nucleus has a radius of 2 µm, and 58.8 s if the nucleus has a radius of 10 µm. Considerably less time is required for an mRNA to travel from a random position within the nucleus to a nuclear pore (0.6 and 40.0 s for nuclei with 2 and 10 µm of radius, respectively). The relation between time for an mRNP to reach a pore and intranuclear position of the gene template is depicted in Fig. 6C. Taking into account that in HeLa cells the total number of pores per nucleus is approximately 2000 and 4000 for G1 and G2 cells, respectively (Maul et al. 1972), we asked what is the effect of increasing the number of pores on the time taken by an mRNP to reach a pore. The simulation results reveal that time is not significantly altered when the total number of pores varies between 2000 and 4000 (Fig. 6D), suggesting that changing the number of nuclear pores within the physiological range will not play a major role in regulation of nucleocytoplasmic transport. Therefore, the physiological number of pores present in a nucleus appears to be quite a robust parameter for optimized export efficiency.

In a nucleus with a 5-µm radius and 2000 pores, an mRNP diffusing at $0.6 \, \mu m^2 \, s^{-1}$ will take on average 6 s to move from a random position in the nucleus to a nuclear pore. In the worst scenario (i.e., an mRNP synthesized in the center of the nucleus), the time to reach a pore is about 10 s. These time periods are not significantly affected by the presence of large intranuclear obstacles such as nucleoli, apparently because the block to diffusion imposed by these obstacles is compensated by the resulting decrease in volume through which the particles must travel. In conclusion, our simulation results reveal that mRNPs can move to the nuclear pore in much less time than the previously anticipated 30 s (Politz et al. 1999).

6
Concluding Remarks

Several lines of recent evidence indicate that mRNAs move about freely in the nonchromosomal space of the nucleus (Politz et al. 1999; Singh et al. 1999), and the simulation results presented here suggest that these molecules can reach a nuclear pore on average in 6 s. Therefore, it appears unlikely that motor-driven transport mechanisms with inherent directionality operate inside the nucleus to move newly transcribed mRNAs to the pores. However, the nucleus is not a uniform fluid medium, and consequently, diffusion is impaired by obstructions imposed by macromolecular crowding. Consistent with this view, large macromolecules such as 2000-kDa dextrans are partially immobilized in the nucleus (Seksek et al. 1997). Conversely, mRNPs (which are expected to be even larger than 2000 kDa) are completely mobile inside the nucleus (Calapez et al. 2002). When cells are depleted of ATP or incubated at reduced temperature, the mobility of mRNPs is significantly reduced, suggesting that energy consumption is required to counteract obstructions to diffusion of these very large complexes (Calapez et al. 2002). Additional reports further suggest that ATP-dependent processes influence the mobility of nuclear bodies and chromatin (reviewed in Carmo-Fonseca et al. 2002). Thus, it is possible that the mechanisms underlying movement of macromolecules including mRNP complexes in the nucleus involve a combination of passive diffusion and energy-dependent reactions.

Acknowledgments. The authors acknowledge Tom Misteli for advice on the manuscript. This work was supported by Fundação para a Ciência e Tecnologia (FCT), Portugal. J.B. and J.R. are recipients of FCT fellowships.

References

Axelrod D, Koppel DE, Schlessinger J, Elson E, Webb WW (1976) Mobility measurement by analysis of fluorescence photobleaching recovery kinetics. Biophys J 16:1055–1069

Bachi A, Braun IC, Rodrigues JP, Panté N, Ribbeck K, von Kobbe C, Kutay U, Wilm M, Görlich D, Carmo-Fonseca M, Izaurralde E (2000) The C-terminal domain of TAP interacts with the nuclear pore complex and promotes export of specific CTE-bearing RNA substrates. RNA 6:136–158

Bauren G, Belikov S, Wieslander L (1998) Transcriptional termination in the Balbiani ring 1 gene is closely coupled to 3'-end formation and excision of the 3'-terminal intron. Genes Dev 12:2759–2769

Braun IC, Herold A, Rode M, Conti E, Izaurralde E (2001) Overexpression of TAP/p15 heterodimers bypasses nuclear retention and stimulates nuclear mRNA export. J Biol Chem 276:20536–20543

Calado A, Carmo-Fonseca M (2000) Localization of poly(A)-binding protein 2 (PABP2) in nuclear speckles is independent of import into the nucleus and requires binding to poly(A)RNA. J Cell Sci 113:2309–2318

Calado A, Kutay U, Kühn U, Wahle E, Carmo-Fonseca M (2000) Deciphering the cellular pathway for transport of poly(A)-binding protein II. RNA 6:245–256

Calapez A, Pereira HM, Calado A, Braga J, Rino J, Carvalho C, Tavanez JP, Wahle E, Rosa AC, Carmo-Fonseca M (2002) The intranuclear mobility of messenger RNA binding proteins is ATP dependent and temperature sensitive. J Cell Biol 159:795–805

Carmo-Fonseca M, Platani M, Swedlow JR (2002) Macromolecular mobility inside the cell nucleus. Trends Cell Biol 12:491–495

Conti E, Izaurralde E (2001) Nucleocytoplasmic transport enters the atomic age. Curr Opin Cell Biol 13:310–319

Dreyfuss G, Kim VN, Kataoka N (2002) Messenger-RNA-binding proteins and the messages they carry. Nat Rev Mol Cell Biol 3:195–205

Ellenberg J, Lippincott-Schwartz J (1999) Dynamics and mobility of nuclear envelope proteins in interphase and mitotic cells revealed by green fluorescence protein chimeras. Methods 19:362–372

Feldherr CM, Kallenbach E, Schultz N (1984) Movement of karyophilic protein through the nuclear pores of oocytes. J Cell Biol 99:2216–2222

Görlich D, Kutay U (1999) Transport between the cell nucleus and the cytoplasm. Annu Rev Cell Dev Biol 15:607–660

Houtsmuller AB, Vermeulen W (2001) Macromolecular dynamics in living cell nuclei revealed by fluorescence redistribution after photobleaching. Histochem. Cell Biol 115:13–21

Kimura H, Sugaya K, Cook PR (2002) The transcription cycle of RNA polymerase II in living cells. J Cell Biol 159:777–782

Klonis N, Rug M, Harper I, Wickham M, Cowman A, Tilley L (2002) Fluorescence photobleaching analysis for the study of cellular dynamics. Eur Biophys J 31:36–51

Kubitscheck U, Wedekind P, Zeidler O, Grote M, Peters R (1996) Single nuclear pores visualized by confocal microscopy and image processing. Biophys J 70:2067–2077

Martins J, Naqvi KR, Melo E (2000) Kinetics of two-dimensional diffusion-controlled reactions: a Monte Carlo simulation of hard disk reactants undergoing a Pearson-type random walk. J Phys Chem 104:4986–4991

Maul GG, Maul HM, Scogna JE, Lieberman MW, Stein GS, Hsu BY, Borun TW (1972) Time sequence of nuclear pore formation in phytohemagglutinin-stimulated lymphocytes and in HeLa cells during the cell cycle. J Cell Biol 55:433–447

Phair RD, Misteli T (2000) High mobility of proteins in the mammalian cell nucleus. Nature 404:604–609

Phair RD, Misteli T (2001) Kinetic modeling approaches to in vivo imaging. Nat Rev Mol Cell Biol 2:898–907

Politz JC, Tuft RA, Pederson T, Singer R (1999) Movement of nuclear poly(A) RNA throughout the interchromatin space in living cells. Curr Biol 9:285–291

Proudfoot NJ, Furger A, Dye MJ (2002) Integrating mRNA processing with transcription. Cell 108:501–512

Reed R, Hurt E (2002) A conserved mRNA export machinery coupled to pre-mRNA splicing. Cell 108:523–531

Reits EAJ, Neefjes JJ (2001) From fixed to FRAP: measuring protein mobility and activity in living cells. Nature Cell Biol 3:E145–E147

Ribbeck K, Görlich D (2001) Kinetic analysis of translocation through nuclear pore complexes. EMBO J 6:1320–1330

Seksek O, Biwersi J, Verkam AS (1997) Translational diffusion of macromolecule-sized solutes in cytoplasm and nucleus. J Cell Biol 138:131–142

Singh OP, Björkroth B, Masich S, Wieslander L, Daneholt B (1999) The intranuclear movement of Balbiani ring pre-messenger ribonucleoprotein particles. Exp Cell Res 251:135–146

Wahle E (1991) A novel poly(A)-binding protein acts as a specificity factor in the second phase of messenger RNA polyadenylation. Cell 66:759–768

Weis K (2002) Nucleocytoplasmic transport: cargo trafficking across the border. Curr Opin Cell Biol 14:328–335

White J, Stelzer E (1999) Photobleaching GFP reveals protein dynamics inside live cells. Trends Cell Biol 9:61–65

Imaging of Single mRNAs in the Cytoplasm of Living Cells

Dahlene Fusco[1], Edouard Bertrand[2] and Robert H. Singer[1]

1
Rationale for Live Cell Imaging of mRNA

Recent advances in mRNA visualization technology have placed biomedical imaging at an interface between molecular biology and cellular biology. It is now increasingly possible to study in vivo gene expression at the transcript level. Analyses of mRNA expression, movement, interactions, and localization will enhance understanding of cellular responses to various conditions and will complement studies of protein expression.

The ability of mRNAs to move in the cytoplasm of eukaryotic cells is essential for sorting sites of synthesis for specific proteins. Studies analyzing the mechanisms of mRNA localization have demonstrated the existence of complex mRNA transport systems. The cytoskeleton is usually required for mRNA localization, and, in a few instances, molecular motors have been shown to be associated with localized mRNA and are required for their movements (Bertrand et al. 1998; Schnorrer et al. 2000; Cha et al. 2001; Wilkie and Davis 2001). In yeast, *ASH1* mRNPs are directly associated with a specific myosin motor, and their movements require both this motor and actin filaments (Bertrand et al. 1998). In the *Drosophila* blastocyst, apically localized mRNAs are thought to be associated with dynein, since their movements require both dynein and microtubules (Schnorrer et al. 2000; Cha et al. 2001). These and other studies have provided strong support for the idea that localized mRNAs are actively transported on cytoskeletal filaments.

In contrast to localized mRNAs, nonlocalized ones are thought to be distributed homogeneously throughout the cytoplasm. Nevertheless, they could still be capable of movement. For instance, following nuclear export, they must leave the perinuclear area to be translated throughout the cytoplasm. Studies with inert tracers, such as fluorescent dextrans or ficolls, suggested that particles with sizes equivalent to some mRNPs have limited diffusion in the cytoplasm (Luby-Phelps et al. 1987). Therefore, even nonlocalized mRNA molecules may require an active transport process. Transport of localized and

[1]Department of Anatomy and Structural Biology and Cell Biology, Albert Einstein College of Medicine, 10461, Bronx, New York, USA, e-mail: rhsinger@aecom.yu.edu
[2]Institut de Genetique Moleculaire de Montpellier-CNRS, UMR 5535, IFR 24, 1919 route de Mende, 34293, Montpellier Cedex 5, France

Progress in Molecular and Subcellular Biology
P. Jeanteur (Ed.): RNA Trafficking and Nuclear Structure Dynamics
© Springer-Verlag Berlin Heidelberg 2003

nonlocalized mRNA should differ, to eventually generate their particular distributions. Thus, study of mRNA transport processes will elucidate behavioral differences among mRNA subpopulations.

Furthermore, live cell mRNA imaging technology may add new information about where and when transcription and translation occur, and can provide a single cell expression profile that cannot be achieved with microchip or biochemical analyses. Simultaneous analysis of multiple transcription sites can provide a single cell profile of gene expression that can be linked precisely to cellular morphology (Levsky et al. 2002) and observation of these transcription sites in living cells will provide novel information about the time course of gene expression in normal and pathological samples.

There are potential medical applications for mRNA imaging, extending from studies of disease mechanism to diagnostic and treatment applications. One of the most significant contributions of mRNA studies to the identification of a disease mechanism was the analysis of trinucleotide repeat transcript foci in nuclei of myotonic dystrophy cells and tissues (Taneja et al. 1995; Davis et al. 1997). This study revealed that myotonic dystrophy (DM) protein kinase transcripts containing increased numbers of CTG repeats accumulate in the nucleus in stable, long-lived clusters. Nuclear accumulation of mutant transcripts is consistent with a model of DM loss of function due to lack of nuclear export, along with possible dominant-negative effects on the export of other mRNA export. Another disease mechanism that can be studied using mRNA visualization is the process of metastasis. Localization patterns of certain mRNAs such as β-actin appear to be related to metastatic capability; nonmetastatic tumor cells differ in their mRNA localization and motility characteristics compared to metastatic tumor cells (Shestakova et al. 1999). Analyses of mRNAs are also being applied to the study of Fragile X Mental Retardation, in which the RNA binding protein, fragile X mental retardation protein (FMRP), is disrupted, providing a disconnect between mRNA regulation and normal neuronal development (Ashley et al. 1993; Siomi et al. 1993). Translational dysregulation of mRNAs normally associated with FMRP may be the proximal cause of fragile X syndrome, and one proposed model for FMRP function is that it shuttles mRNAs from the nucleus to postsynaptic sites where transcripts are held in a translationally inactive form until further stimulation alters FMRP function (Brown et al. 2001). Several other mRNA binding proteins have been shown to play a role in mRNA trafficking in developing neurons. These neuronal mRNA binding proteins include zipcode binding proteins one and two (ZBP-1, ZBP-2; Ross et al. 1997; Gu et al. 2002), which play a role in the dendritic trafficking of β actin mRNA. Live cell mRNA studies can also provide important insights in virology. Subcellular localization and tracking of viral sequences, combined with labeling of relevant proteins and other structures, will elucidate viral host interaction mechanisms. These studies will also help clarify the cellular lifecycle of viral RNA as well as host cell defense mechanism and may correlate cell phenotype with viral RNA load as infection progresses.

Messenger RNA detection in individual, living cells has several advantages over other techniques for the study of gene expression. For example, in comparison to population analyses, such as Northern blotting, single cell studies can describe the amount of virus in individual infected cells, or the percentage of cells that are infected. By imaging mRNA in living cells instead of fixed cells, it is possible to determine the timing of mRNA expression and dynamics, as well as different patterns of mRNA movement and interactions throughout the lifetime of a transcript. Because mRNA movement appears to be a rapid, rare event, the ability to examine a series of time points becomes important for the observation of mRNA transit. In addition, live cell imaging avoids the possible introduction of artifact during cell fixation.

For years, the major limitation of imaging mRNA in living cells has been a dearth of available technology. Recent years have yielded significant advances in four major categories of mRNA visualization technology, including microinjection of fluorescent RNA (Ainger et al. 1993; Ferrandon et al. 1994), hybridization of fluorescent oligonucleotide probes (Politz et al. 1998, 1999), the use of cell-permeant dyes which bind RNA, and sequence-specific mRNA recognition by a GFP fusion protein (Bertrand et al. 1998). Each of these technologies (Fig. 1, by permission) will be presented, but the focus of this article will be the sequence-specific recognition of mRNA by GFP fused to the mRNA binding protein MS2.

2
Existing Technologies for the Visualization of RNA in Living Cells

2.1
Microinjection of Fluorescent RNA

Fluorescent RNA can be synthesized by in vitro transcription with phage polymerases in the presence of fluorescent nucleotide analogs for microinjection into living cells (Wilkie and Davis 2001). Likewise, transcription with a mixture of unmodified and amino-allyl nucleotides generates RNA that can be chemically coupled to activated fluorophores (Wang et al. 1991). Because amino-allyl nucleotides are incorporated at a higher frequency than fluorescent analogs, very bright RNA can be obtained. The main disadvantage of microinjecting fluorescently labeled mRNA is the introduction of nonendogenous material into the cell. Nevertheless, this approach has proven successful in a number of cases to reveal routes of trafficking of localized RNAs (Ainger et al. 1993; Theurkauf and Hazelrigg 1998; Wilkie and Davis 2001).

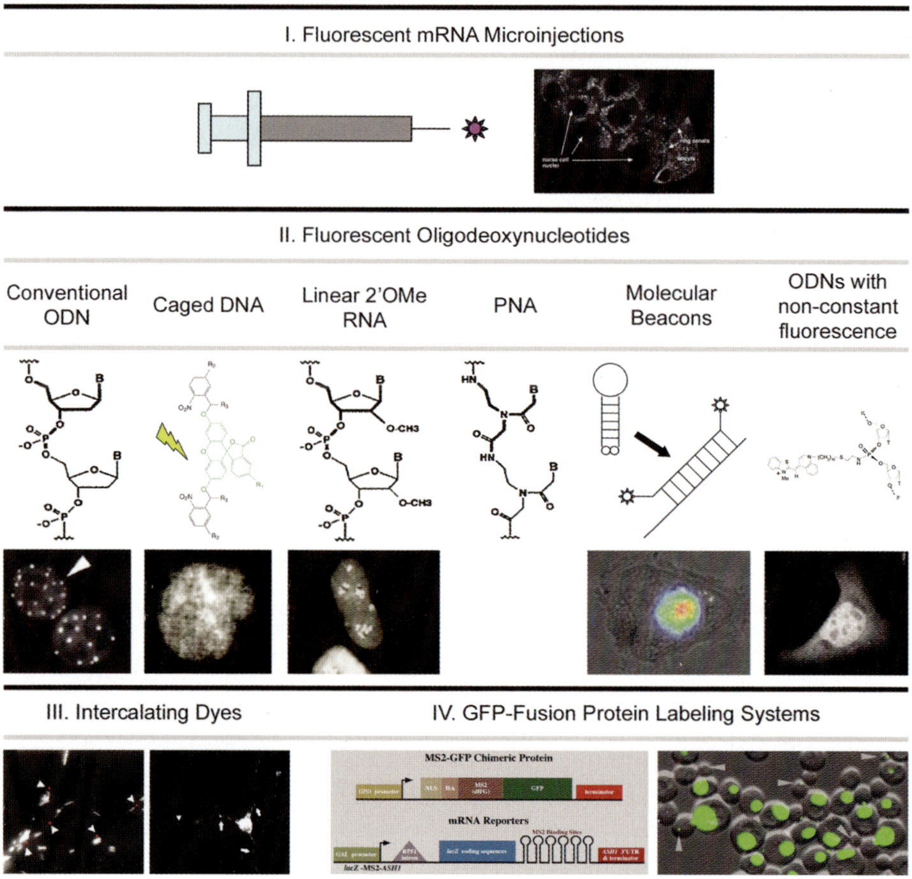
I. Fluorescent mRNA Microinjections
II. Fluorescent Oligodeoxynucleotides
Conventional ODN
Caged DNA
Linear 2'OMe RNA
PNA
Molecular Beacons
ODNs with non-constant fluorescence
III. Intercalating Dyes
IV. GFP-Fusion Protein Labeling Systems
MS2-GFP Chimeric Protein
mRNA Reporters

Fig. 1. Existing technology for the visualization of mRNA in living cells, including images obtained using those technologies (Knowles et al. 1996, 1997; Bertrand et al. 1998; Theurkauf et al. 1998; Politz et al. 1999; Lorenz et al. 2000; Dirks et al. 2001; Perlette et al. 2001; Privat et al. 2001)

2.2 Fluorescent Oligonucleotide Probes (FIVH-Fluorescent In Vivo Hybridization)

Fluorescent oligonucleotide probes have been developed to follow unmodified, endogenous cellular RNA. An oligonucleotide complementary to the desired sequence is covalently linked to fluorochromes and can contain chemical modifications that increase cell penetration and hybrid stability. Oligonucleotides are delivered across the plasma membrane and hybridize with the target sequence to identify RNA directly in living cells (Politz et al. 1995, 1999). Six categories of fluorescent oligonucleotide probes exist, including conventional DNA probes (Politz et al. 1995; Lorenz et al. 2000), caged DNA probes (Politz et al. 1999; Pederson 2001), linear 2' O methyl (2'OMe) RNA probes (Carmo-Fonseca et al. 1991; Dirks et al. 2001; Molenaar et al. 2001), peptide nucleic acid (PNA) probes (Dirks et al. 2001), molecular beacons (Tyagi et al. 1996, 1998; Sokol et al. 1998; Molenaar et al. 2001), and probes with a linked fluorophore whose fluorescence properties change upon association with the target mRNA strand (Thuong et al. 1987; Privat et al. 2001). Advantages of oligodeoxynucleotide (ODN) probes are their ability to detect endogenous mRNA, their relative stability (as long as 18 h, Politz et al. 1995), and the ease with which they are introduced into the cell. ODNs can be used to perform fluorescence resonance energy transfer (FRET) by simultaneously introducing two oligonucleotides labeled with different fluorophores and recording the loss of donor fluorescence and/or gain of acceptor fluorescence. A caveat of antisense oligodeoxyribonucleotides is that they can prevent translation by blocking ribosome movement along the mRNA or by inducing RNase H cleavage of the RNA/oligo hybrid, altering expression of target mRNA. In addition, fluorescent oligonucleotides have been found to localize exclusively to the nucleus in some cases (Dirks et al. 2001).

2.3
Cell Permeant Dyes

SYTO-14 is a membrane-permeable nucleic acid stain that has been used to study RNA in living cells (Knowles et al. 1996, 1997). SYTO-14 is advantageous because it will detect the endogenous RNA population, it is not limited to the nucleus, and its use requires no subsequent intervention after introduction into cells. A major disadvantage of cell permeant RNA binding dyes is their lack of specificity due to identical recognition of all RNAs.

2.4
RNA Tagged with Sequence-Specific GFP Fusion Proteins

It is possible to visualize a specific transcript by fusion of GFP with a sequence-specific RNA binding protein, such as the coat protein of phage MS2, and insertion of the RNA sequence recognized by this protein in the RNA of interest. When both the MS2-GFP fusion protein and the reporter RNA are co-expressed, MS2-GFP binds to the reporter in living cells and GFP fluorescence constitutes a reliable indicator of RNA localization (Bertrand et al. 1998). To remove the noise of unbound MS2-GFP, it is possible either to express a low level of MS2-GFP such that most of the MS2-GFP molecules are bound to the RNA, or to include a localization signal in the MS2-GFP protein which will exclude unbound molecules from the area of interest (Bertrand et al. 1998). This technique has several advantages: (1) it is possible to visualize RNA that is synthesized and processed by the cell; (2) it is possible to use the technique in organisms that cannot be microinjected, or that do not allow penetration of macromolecules; (3) it is possible to increase the sensitivity of RNA detection by multimerizing the binding site in the RNA reporter. The detection limit of this technique has reached the level of single mRNA molecules through the use of MS2 repeats containing 6 to 24 binding sites (i.e., 12–48 GFP molecules; Fusco et al. 2003). An analogous detection system has been designed to visualize genes in living nuclei, using a lac i-GFP fusion protein and an array of lac o binding sites inserted in the genome. This system first demonstrated the limits of detection, and requires as little as 60 GFP molecules (Robinett et al. 1996). The combination of these two systems, for the simultaneous visualization of gene activation and mRNA trafficking in living cells, is in progress.

However, the MS2-GFP system does involve the visualization of a nonendogenous mRNA, introduces a nonendogenous protein into the RNA movement process, and does not allow cellular control of expression levels. While the first two disadvantages are inherent to the MS2 labeling system, the third disadvantage can be addressed by design of inducible cell lines with specific control of MS2 reporter levels. Alternatively, an endogenous GFP-labeled RNA binding protein can be used. Theurkauf and Hazelrigg tagged bicoid mRNA with a GFP-exuperantia fusion (Theurkauf et al. 1998). While this system avoids introduction of an exogenous protein other than GFP, it is difficult to determine whether a protein that is known to bind an mRNA is, in fact, persistently bound to that mRNA, thus presenting the question of whether all GFP fusion protein movements are indicative of mRNA movements. The MS2-GFP-labeling system provides a stable link between GFP and mRNA (K_d = 39nM, in vitro; Lago et al. 1998), allowing reliable identification of mRNA particles throughout trafficking behavior. Importantly, when the fusion protein contains a nuclear localization signal (NLS), it will be targeted to the nucleus and only be introduced into the cytoplasm by its association with an exported RNA. Finally, the multimerization of the MS2-GFP allows accumulation of a fluorescent signal sufficient to characterize single molecules.

3
Analysis of Single RNA Dynamics in Living Cells

Following successful visualization of the RNA of interest, one must choose a method for analysis of RNA dynamics. Two general techniques have been used: photobleaching and fluorescent correlation spectroscopy. Photobleaching techniques involve irradiation with intense light leading to the irreversible inactivation of fluorophores (Axelrod et al. 1976; Phair and Misteli 2000). Fluorescence correlation spectroscopy (FCS) is a fluctuation analysis method used to measure molecular transport and chemical kinetics by counting the molecules entering and leaving a small interrogation volume over successive time intervals (Wang et al. 1991). When single RNA molecules are detectable, direct tracking of individual RNA by time-lapse microscopy is applicable. These results yield a population profile of RNA movements.

4
Time-Lapse Microscopy/Particle Tracking

It therefore becomes possible to use time-lapse microscopy to follow RNA movements. This requires either single molecule sensitivity, or the concentration of many RNA molecules in a multi-molecule aggregate, as previously observed in some cases (Ainger et al. 1993; Ferrandon et al. 1994; Bertrand et al. 1998; Wilkie et al. 2001). The single molecule dynamics measured in this method are complementary to the population dynamics obtained through fluorescence recovery after photobleaching (FRAP) or fluorescence correlation spectroscopy (FCS). Diffusion coefficients can be obtained from time-lapse data although analysis of a large number of molecules is required, but in addition, the temporal sequence of movements can provide novel information. For example, if a molecule is alternatively immobile and mobile on a relatively short time-scale, photobleaching techniques may average these behaviors while both will be apparent by time-lapse microscopy. Likewise, FCS would only see a small window of the behavior profile of single molecules. The ability to follow the fate of a molecule can reveal an ordered sequence of events. In addition, time-lapse techniques can reveal rare, but functionally important behaviors that could not be detected otherwise. For instance, *ASH1* mRNA visualized in the cytoplasm of living yeast (Bertrand et al. 1998) is localized at the bud tip, after transport by a specific myosin motor on actin cables. However, because the transport is very rapid and lasts for less than a minute of the entire cellular lifetime, the proportion of actively transported mRNA molecules is on average very low and does not exceed a few percent of the total molecules. Such a minor subset of behavior in a population would very likely not be detected by photobleaching and FCS techniques.

When single molecule resolution is achieved, a detailed analysis of individual particle movements can yield information related to the mechanism of mRNA transport. Analysis includes collection of position, velocity, and acceleration values for the determination of movement type, such as active transport, Brownian motion, or corralled diffusion. In the case of active transport, movement analysis can help identify candidate motors involved in transport. For example, if a particle is observed to display actin dependent movement at a constant speed of approximately 200–400 nm/sec, a myosin motor is likely to be involved (Cheney et al. 1993). Other interesting characteristics of mRNA particles that can be obtained from time-lapse images in living cells include particle size, location, diffusion coefficient, mean squared displacement, and relation to other cellular components, such as the cytoskeleton, all of which can be useful in determining the mechanism of mRNA movement and localization. When images are of a high quality, but particles can still not be tracked, it is possible to perform two-dimensional deconvolutions on movies to further improve signal to noise ratio and facilitate particle tracking (Huygens, Bitplane, The Netherlands).

We summarize below the results obtained using the described imaging techniques to study cytoplasmic mRNAs. In the cytoplasm, an increasing number of specific transcripts are being studied in living cells. These studies are significant because they provide unique information about the dynamics of mRNA movement and help elucidate the mechanisms of mRNA particle assembly, transport, and localization as well as crucial mRNA interactions with proteins and the cytoskeleton. This information can be obtained in single cells for endogenous, engineered and viral transcripts, greatly widening our current understanding of mRNA behavior.

5
In Vivo mRNA Analyses in the Cytoplasm

In yeast, *ASH1* mRNA was visualized in live *Saccharomyces cerevisiae*, in the first use of the MS2-GFP labeling system. The GFP- *ASH1* particle required specific *ASH1* sequences for formation and localization to the bud tip (Bertrand et al. 1998). Mutations in the *ASH1* -associated She proteins were found to disrupt particle localization, and *She2* and *She3* mutants also inhibited particle formation. Video microscopy showed that She1p/Myo4p moved mRNA particles to the bud tip at velocities of 200–440 nm/sec. Further studies of the *ASH1* mRNA transport process (Beach et al. 1999) showed that in cells lacking Bud6p/Aip3p or She5p/Bni1p, *ASH1* mRNA particles traveled to the bud, but failed to remain at the bud tip, revealing a distinction between mRNA transport and localization similar to that described in oligodendrocytes (Ainger et al. 1997). These findings not only characterized the movement of the *ASH1* mRNA particle, but also contributed significantly to the identification of proteins, including a motor, that are required for *ASH1* particle transport.

One method for the visualization of mRNA in living cells using GFP is the use of GFP-tagged RNA binding proteins. The dynamic behavior of *bicoid* mRNA in living Drosophila oocytes was visualized with such a mechanism using a GFP-exuperantia fusion (Theurkauf et al. 1998). The results of this study elaborate the relationship between mRNA transport and the cytoskeleton, demonstrating that particle movement can alternate through phases of cytoskeleton dependence and independence.

Additional information regarding the *bicoid* localization mechanism came from microinjection of fluorescent *bicoid* transcripts into Drosophila oocytes (Cha et al. 2001). Interestingly, fluorescent *bicoid* mRNA injected into the oocyte displays nonpolar microtubule-dependent transport to the closest cortical surface, but *bicoid* mRNA injected into the nurse cell cytoplasm, withdrawn, and injected into a second oocyte shows microtubule-dependent transport to the anterior cortex. This study identified the requirement of a nurse cell cytoplasmic component for proper trafficking of *bicoid* mRNA. It is interesting to note that a component that is present in the nurse cell cytoplasm allows reconnection of *bicoid* mRNA to the microtubule network after a microtubule independent transit through the ring canal, indicating that transcripts bear some mark, possibly a bound protein, indicating their prior location even as they continue transit into other compartments. Thus, transcripts in a given compartment may be able to be distinguished by their past behavior based on their present set of interaction partners. This concept of transcript historesis has also been proposed with respect to the behavior of CaMKII granules (Rook et al. 2000). In this case, it was suggested that granule location at a specific point in time reflected the activation history of a synapse, allowing identification of stimuli that were in the past.

Injection of fluorescently labeled *wingless* and pair-rule transcripts into living Drosophila syncytial blastoderm embryos (Wilkie et al. 2001) showed that fluorescent apical RNAs specifically assemble into particles that approach the minus end of microtubules using the motor protein cytoplasmic dynein and its associated dynactin complex.

In neurons, the movement of endogenous RNA was directly visualized in neuronal processes of living cells using the intercalating dye SYTO 14 (Knowles et al. 1996, 1997). One of the first studies in which fluorescently labeled mRNA was microinjected into the cytoplasm of living cells was performed by Ainger et al. (1993), who microinjected myelin basic protein (MBP) mRNA into neurites and observed the formation of granules which were transported along oligodendrocytes and localized to the myelin compartment. More recent analyses of A2RE RNA granule movements, performed with rapid confocal line scanning through a single granule on a microtubule, demonstrated a rapid pattern of back and forth vibrations along the microtubule axis. Through mean squared displacement analysis of these vibrations, it was observed that A2RE granules undergo "corralled diffusion", or movement of a defined distance before reversal of direction (Carson et al. 2001). The movements of GFP-labeled CAMKII mRNA particles were observed in rat hippocampal neurons, in the first use of the MS2-GFP system in somatic cells. Interestingly, this study showed that anterograde particle movements increased in response to neuronal depolarization, in comparison to oscillatory and retrograde movements (Rook et al. 2000). Examination of GFP-labeled β actin zipcode binding protein 1 (ZBP-1) provided insights regarding the translocation of β actin mRNA in neurons (Zhang et al. 2001). Live cell imaging of EGFP-ZBP1 revealed that GFP-labeled particles move in a rapid, bidirectional fashion that is reduced tenfold by antisense oligonucleotides to the β actin zipcode.

In cell lines, the behavior of transiently expressed c-fos mRNA was observed in living Cos-7 cells through microinjection of fluorescently labeled oligonucleotides (Tsuji et al. 2001). Microinjection of molecular beacons targeted to the *vav* protooncogene in K562 human leukemia cells followed by confocal microscopy within 15 min of microinjection demonstrated that molecular beacons could detect transcripts in the cytoplasm of living cells, with a detection limit of ~10 molecules of mRNA (Sokol et al. 1998). A later study showed that most beacon hybridization occurred within 10 min of microinjection and maximum intensity was obtained by 15 min and maintained until 40 min after microinjection. (Perlette and Tan 2001).

To better understand the mechanisms controlling mRNA movements, we developed a method for visualizing single RNA molecules in living mammalian cells in real time. This method uses the MS2-GFP fusion protein and a reporter mRNA containing tandemly repeated MS2 binding sites, inserted between LacZ and SV40 sequences (Bertrand et al. 1998), followed by quantitative in situ hybridization. When the fusion protein was coexpressed with the reporter mRNA, the MS2-GFP bound to the mRNA was exported to the cytoplasm. When the MS2-GFP fusion protein was expressed alone, a nuclear localization signal confined the protein to the nucleus. We used Cos cells in this study because of their optical properties.

Discrete particles of GFP-labeled mRNA were resolvable only when 24 MS2 sites were in the reporter mRNA, not when either 6 or 12 MS2 sites were present. To quantify the number of mRNA molecules in each GFP-labeled particle, we imaged concentrations of purified GFP to determine the amount of light emitted by a given number of GFP molecules, and we calculated the number of GFP molecules in each GFP-labeled mRNA particle (Fusco et al. 2003). Quantification of the fluorescence from more than 600 GFP-labeled mRNA particles in eight cells showed that most GFP-labeled mRNA particles contained 20–50 molecules of GFP, with an average of 33. Since the MS2 protein binds as a dimer (Valegard et al. 1994), this measurement coincided with the expected number of GFP molecules labeling a single molecule of reporter RNA, assuming that most of the 24 MS2 sites were bound. This quantification procedure was independently validated by using a laco/GFPlaci system for calibration of the GFP signal (Robinett et al. 1996). To exclude the possibility that clusters of mRNA molecules sparsely labeled with GFP were misidentified as single mRNA molecules, we performed in situ hybridization to a single target on the mRNA reporter (Femino et al. 1998). Cells were simultaneously hybridized with a Cy5-labeled probe to a region between MS2 binding sites and a single Cy3-labeled probe to the LacZ portion of the mRNA. GFP, Cy5, and Cy3 were detected concurrently in single cells, allowing quantification of the number of probes colocalized with individual GFP particles (Fusco et al. 2003). Analysis of 125 multicolor particles in five cells showed that the majority of multicolored particles were single molecules, containing one Cy3 probe and eight MS2 probes. This validates the independent conclusion, from the analysis described above that the majority of the GFP-labeled mRNA particles detected in the cytoplasm contained a single RNA molecule. There were a small number of GFP particles containing more than a single RNA molecule present in every cell. The single-molecule GFP particles were distinct from the RNA "granules" observed in a variety of cell types, which are much brighter and larger and, although not quantitated, apparently contain multiple RNA molecules (Davis and Ish-Horowicz 1991; Ainger et al. 1993; Bertrand et al. 1998; Theurkauf and Hazelrigg 1998; Rook et al. 2000; Wilkie and Davis 2001; Kloc et al. 2002; Farina et al. 2003).

Figure 2 shows the movements of the single mRNA particles recorded in living cells at a rate of nine images per second. This high frequency was required because some mRNA molecules exceeded 1 µm/s (see below). The mRNAs contained the coding sequence of LacZ, the MS2 sites, and the 3′ UTR of either the human growth hormone gene or SV40 (reporters LacZ-24-hGH and LacZ-24-SV, respectively).

Dynamic behavior of β actin 3′ UTR mRNA was compared to that of human growth hormone (hGH) and SV40 transcripts in Cos7 cells. These three transcripts were chosen because β actin mRNA, hGH mRNA, and SV40 mRNA represent transcripts of three distinct categories: endogenous sequences that have been observed to localize, endogenous sequences that have not been observed to localize, and nonendogenous sequences that have not been observed to localize, respectively. Regardless of any specific cytoplasmic distribution, individual mRNA molecules exhibit rapid and directional movements on microtubules. In addition to directional movements, all three transcript types displayed static, diffusional, and corralled diffusional behavior. Disruption of the cytoskeleton with drugs showed that microtubules and microfilaments are involved in the types of mRNA movements we have observed, which included complete immobility and corralled and nonrestricted diffusion. Individual mRNA molecules switched frequently among these movements, suggesting that mRNAs undergo continuous cycles of anchoring, diffusion, and active transport. Interestingly, transcripts containing the β actin 3′ UTR were observed to undergo directed movement more often than SV40 and hGH transcripts, indicating that the mechanism for mRNA localization may involve a sequence-dependent increase in motor-like movements. While Cos7 cells do not themselves localize β actin mRNA, they do contain the β actin 3′ UTR binding protein, ZBP-1, which may be sufficient to enable the primary steps of localization and the observed increase in sequence-dependent directed movements.

To better determine the behavioral characteristics that underlie mRNA localization, it is necessary to study this process directly in primary cells, such as fibroblasts or neurons, where localization is known to occur. These studies are difficult because of the low transfection efficiency of primary cells and the thickness of their cytoplasm, which complicates particle tracking. Nevertheless, such studies are in progress (Fusco, unpubl. data) and it is hoped that they will yield significant insights into the mechanism of mRNA localization in somatic cells, along with the role of this process in wound healing, metastasis, and neuronal development.

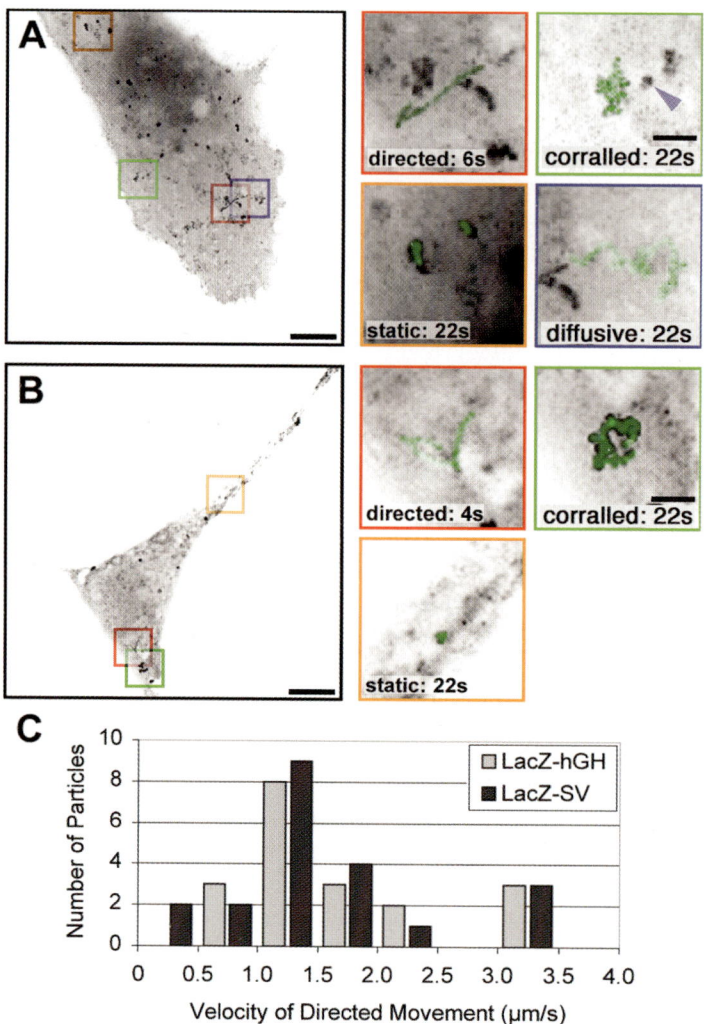

Fig. 2. Dynamics of mRNA molecules in the cytoplasm of mammalian cells (reproduced from Fusco et al. 2003). All images were obtained at a rate of nine images per second, for periods of 22 s, and were deconvolved. **A** Movements of LacZ-24-hGH mRNAs. Cos cells transiently expressing LacZ-24-hGH mRNAs and the MS2-GFP protein were imaged live. *Left* a maximum intensity image projection of 200 time frames on one image ("maximal projection"). The *scale bar* represents 10 µm. *Right* panel magnifications: the *scale bar* represents 2 µm mRNA track superimposed (*green*) from each of the indicated boxed regions. The *blue arrow* points to a "static" particle in the vicinity of a "corralled" mRNA. **B** Movements of LacZ-24-SV mRNAs. Cos cells were transiently cotransfected with pRSV-Z-24-SV and pMS2-GFP and were imaged as in **A**. The *scale bar* represents 10 µm. *Right* panel magnifications: track of mRNA movement superimposed (*green*) on an enlargement from each of the indicated boxed regions. The scale bar represents 2 µm. **C** Velocities of directed motion. For each directed movement of either the LacZ-24-hGH or the LacZ-24-SV mRNA particles, the mean velocity was calculated. The corresponding histograms show a peak at 1.0–1.5 µm/s (Knowles et al. 1996, 1997; Bertrand et al. 1998; Theurkauf et al. 1998; Politz et al. 1999; Lorenz et al. 2000; Dirks et al. 2001; Perlette et al. 2001; Privat et al. 2001)

6
Conclusion

The field of live cell mRNA analysis is in a period of reorientation, slowly shifting focus from the development of new visualization technology to the application of that technology to important biological and medical questions. Significant advances have been made already, including the detailed characterization of mRNA transport and localization particles in yeast (Bertrand et al. 1998) and in Drosophila oocytes (Theurkauf and Hazelrigg 1998; Cha et al. 2001), signal transduction mechanisms related to mRNA transport in neurons (Rook et al. 2000) and in fibroblasts (Latham et al. 2001), and sequence-dependent movement differences in mammalian cell lines (Fusco et al. 2003). Despite these advances, major questions remain. For example, where are the multi-molecule mRNA "granules" formed, in the nucleus or the cytoplasm? In the cytoplasm, what is the mechanism of mRNA localization, and does it include a sequence-specific process of directed movement, of packaging, of anchoring, or some combination of these processes? Is the mechanism the same for all localized transcripts and what are the signal transduction pathways involved in stimulating localization? Finally, what are the medical consequences of disrupted mRNA trafficking and the medical applications of altered mRNA trafficking? As the technology described above becomes better practised and applied, we will doubtless find answers to many of these questions and advance our understanding of molecular and cellular biology in the process. Furthermore, the future will yield new developments in imaging cells in live animals (Farina et al. 1998) and will eventually unveil the details of the RNA lifecycle.

Acknowledgments. The authors thank Shailesh Shenoy for his help with the figures. Supported by NIH GM 54887.

References

Ainger K, Avossa D, Morgan F, Hill SJ, Barry C, Barbarese E, Carson JH (1993) Transport and localization of exogenous myelin basic protein mRNA microinjected into oligodendrocytes. J Cell Biol 123:431–441

Ainger K, Avossa D, Diana AS, Barry C, Barbarese E, Carson JH (1997) Transport and localization elements in myelin basic protein mRNA. J Cell Biol 138:1077–1087

Ashley CT Jr, Wilkinson KD, Reines D, Warren ST (1993) FMR1 protein: conserved RNP family domains and selective RNA binding. Science 262(5133):563–566

Axelrod D, Koppel DE, Schlessinger J, Elson E, Webb WW (1976) Mobility measurement by analysis of fluorescence photobleaching recovery kinetics. Biophys J 16:1055–1069

Beach DL, Salmon ED, Bloom K (1999) Localization and anchoring of mRNA in budding yeast. Curr Biol 9:569–578

Bertrand E, Chartrand P, Schaefer M, Shenoy SM, Singer RH, Long RM (1998) Localization of ASH1 mRNA particles in living yeast. Mol Cell 2:437–445

Brown V, Jin P, Cernan S, Darnell JC, O'Donnell WT, Tenenbaum SA, Jin X, Feng Y, Wilkinson KD, Keene JD, Darnell RB, Warren ST (2001) Microarray identification of FMRP-associated brain mRNAs and altered mRNA translational profiles in fragile X syndrome. Cell 107:477–487

Carmo-Fonseca M, Pepperkok R, Sproat BS, Ansorge W, Swanson MS, Lamond AI (1991) In vivo detection of snRNP-rich organelles in the nuclei of mammalian cells. EMBO J 10:1863–1873

Carson JH, Cui H, Barbarese E (2001) The balance of power in RNA trafficking. Curr Opin Neurobiol 11:558–563

Cha BJ, Koppetsch BS, Theurkauf WE (2001) In vivo analysis of Drosophila bicoid mRNA localization reveals a novel microtubule-dependent axis specification pathway. Cell 106:35–46

Cheney RE, O'Shea MK, Heuser JE, Coelho MV, Wolenski JS, Espreafico EM, Forscher P, Larson RE, Mooseker MS (1993) Brain myosin-V is a two-headed unconventional myosin with motor activity. Cell 75:13–23

Davis I, Ish-Horowicz D (1991) Apical localization of pairrule transcripts requires 3' sequences and limits protein diffusion in the Drosophilia blastoderm embryo. Cell 67:927–940

Davis BM, McCurrach ME, Taneja KL, Singer RH, Housman DE (1997) Expansion of a CUG trinucleotide repeat in the 3' untranslated region of myotonic dystrophy protein kinase transcripts results in nuclear retention of transcripts. Proc Natl Acad Sci USA 94:7388–7393

Dirks RW, Molenaar C, Tanke HJ (2001) Methods for visualizing RNA processing and transport pathways in living cells. Histochem Cell Biol 115:3–11

Farina KL, Wyckoff JB, Rivera J, Lee H, Segall JE, Condeelis JS, Jones JG (1998) Cell motility of tumor cells visualized in living intact primary tumors using green fluorescent protein. Cancer Res 58:2528–2532

Farina KL, Huettelmaier S, Musunuru K, Darnell R, Singer RH (2003) Two ZBP1 KH domains facilitate β-actin mRNA localization, granule formation, and cytoskeletal attachment. J Cell Biol 160:77–87

Femino AM, Fay FS, Fogarty K, Singer RH (1998) Visualization of single RNA transcripts in situ. Science 280:585–590

Ferrandon D, Elphick L, Nusslein-Volhard C, St Johnston D (1994) Staufen protein associates with the 3' UTR of bicoid mRNA to form particles that move in a microtubule-dependent manner. Cell 79:1221–1232

Fusco D, Accornero N, Lavoie B, Shenoy SM, Blanchard J, Singer RH, Bertrand E (2003) Sequence dependent movement of single mRNA molecules in the cytoplasm of living mammalian cells. Curr Biol 13:161–167

Gu W, Pan F, Zhang H, Bassell GJ, Singer RH (2002) A predominantly nuclear protein affecting cytoplasmic localization of beta-actin mRNA in fibroblasts and neurons. J Cell Biol 156:41–51

Kloc M, Zearfoss NR, Etkin LD (2002) Mechanisms of subcellular mRNA localization. Cell 108:533–544

Knowles RB, Sabry JH, Martone ME, Deerinck RJ, Ellisman MH, Bassell GJ, Kosik KS (1996) Translocation of RNA granules in living neurons. J Neurosci 16:7812–7820

Knowles RB, Kosik KS (1997) Neurotrophin-3 signals redistribute RNA in neurons. Proc Natl Acad Sci USA 94:14804–14808

Lago H, Fonseca SA, Murray JB, Stonehouse NJ, Stockley PG (1998) Dissecting the key recognition features of the MS2 bacteriophage translational repression complex. Nuc Acid Res 26(5):1337–1344

Latham VM, Yu EH, Tullio AN, Adelstein RS, Singer RH (2001) Rho-dependent signaling pathway operating through myosin localizes beta-actin mRNA in fibroblasts. Curr Biol 11:1010–1016

Levsky JM, Shenoy SM, Pezo R, Singer RH (2002) Single-cell gene expression profiling. Science 297(5582):836–840

Lorenz P, Misteli T, Baker BF, Bennett CF, Spector DL (2000) Nucleocytoplasmic shuttling: a novel in vivo property of antisense phosphorothioate oligodeoxynucleotides. Nucleic Acids Res 28:582–592

Luby-Phelps K, Castle PE, Taylor DL, Lanni F (1987) Hindered diffusion of inert tracer particles in the cytoplasm of mouse 3T3 cells. Proc Natl Acad Sci USA 84:4910–4913

Molenaar C, Marras SA, Slats JCM, Truffert J-C, Lemaitre M, Raap AK, Dirks RW, Tanke HJ (2001) Linear 2′ O-Methyl RNA probes for the visualization of RNA in living cells. Nucleic Acids Res 29:E89

Pederson T (2001) Fluorescent RNA cytochemistry: tracking gene transcripts in living cells. Nucleic Acids Res 29:1013–1016

Perlette J, Tan W (2001) Real-time monitoring of intracellular mRNA hybridization inside single living cells. Anal Chem 73:5544–5550

Phair RD, Misteli T (2000) High mobility of proteins in the mammalian cell nucleus. Nature 404:604–609

Politz JC, Taneja KL, Singer RH (1995) Characterization of hybridization between synthetic oligodeoxynucleotides and RNA in living cells. Nucleic Acids Res 23:4946–4953

Politz JC, Browne ES, Wolf DE, Pederson T (1998) Intranuclear diffusion and hybridization state of oligonucleotides measured by fluorescence correlation spectroscopy in living cells. Proc Natl Acad Sci USA 95:6043–6048

Politz JC, Tuft RA, Pederson T, Singer RH (1999) Movement of nuclear poly (A) RNA throughout the interchromatin space in living cells. Curr Biol 9:285–291

Privat E, Melvin T, Asseline U, Vigny P (2001) Oligonucleotide-conjugated thiazole orange probes as "light-up" probes for messenger ribonucleic acid molecules in living cells. Photochem Photobiol 74:532–541

Robinett CC, Straight A, Li G, Willhelm C, Sudlow G, Marray A, Belmont AS (1996) In vivo localization of DNA sequences and visualization of large-scale chromatin organization using lac operator/repressor recognition. J Cell Biol 135:1685–1700

Rook MS, Lu M, Kosik KS (2000) CaMKIIalpha 3′ untranslated region-directed mRNA translocation in living neurons: visualization by GFP linkage. J Neurosci 20:6385–6393

Ross AF, Oleynikov Y, Kislauskis EH, Taneja KL, Singer RH (1997) Characterization of a beta-actin mRNA zipcode-binding protein. Mol Cell Biol 17:2158–2165

Schnorrer F, Bohmann K, Nusslein-Volhard C (2000) The molecular motor dynein is involved in targeting swallow and biocoid RNA to the anterior pole of Drosophila oocytes. Nat Cell Biol 2:185–190

Shestakova EA, Wyckoff J, Jones J, Singer RH, Condeelis J (1999) Correlation of beta-actin messenger RNA localization with metastatic potential in rat adenocarcinoma cell lines. Cancer Res 59:1202–1205

Siomi H, Siomi MC, Nussbaum RL, Dreyfuss G (1993) The protein product of the fragile X gene, FMR1, has characteristics of an RNA-binding protein. Cell 74:291–298

Sokol DL, Zhang X, Lu P, Gewirtz AM (1998) Real time detection of DNA.RNA hybridization in living cells. Proc Natl Acad Sci USA 95:11538–11543

Taneja KL, McCurrach M, Schalling M, Housman D, Singer RH (1995) Foci of trinucleotide repeat transcripts in nuclei of myotonic dystrophy cells and tissues. J Cell Biol 128:995–1002

Theurkauf WE, Hazelrigg TI (1998) In vivo analyses of cytoplasmic transport and cytoskeletal organization during Drosophila oogenesis: characterization of a multi-step anterior localization pathway. Development 125:3655–3666

Thuong NT, Asseline U, Roig V, Takasugi M, Helene C (1987) Oligo (alpha-deoxynucleotide) s covalently linked to intercalating agents: differential binding to ribo- and deoxyribo-polynucleotides and stability towards nuclease digestion. Proc Natl Acad Sci USA 84:5129–5133

Tsuji A, Sato Y, Hirano M, Suga T, Koshimoto H, Taguchi T, Ohsuka S (2001) Development of a time-resolved fluorometric method for observing hybridization in living cells using fluorescence resonance energy transfer. Biophys J 81:501–515

Tyagi S, Kramer FR (1996) Molecular beacons: probes that fluoresce upon hybridization. Nat Biotechnol 14:303–308

Tyagi S, Bratu DP, Kramer FR (1998) Multicolor molecular beacons for allele discrimination. Nat Biotechnol 16:49–53

Valegard K, Murray JB, Stockley PG, Stonehouse NJ, Lijas L (1994) Crystal structure of an RNA bacteriophage coat protein-operator complex. Nature 371:623–626

Wang J, Cao LG, Wang YL, Pederson T (1991) Localization of pre-messenger RNA at discrete nuclear sites. Proc Natl Acad Sci USA 88:7391–7395

Wilkie GS, Davis I (2001) Drosophila wingless and pair-rule transcripts localize apically by dynein-mediated transport of RNA particles. Cell 105:209–219

Zhang HL, Eom T, Oleynikov Y, Shenoy SM, Liebelt DA, Dictenberg JB, Singer RH, Bassell GJ (2001) Neurotrophin-induced transport of a beta-actin mRNP complex increases beta-actin levels and stimulates growth cone motility. Neuron 31:261–275

Subject Index

Printing: Krips bv, Meppel, The Netherlands
Binding: Stürtz, Würzburg, Germany